Rising Powers, People Rising

Rising Powers, People Rising is a pathbreaking volume in which leading international scholars discuss the emerging political economy of development in the BRICS countries centred on neo-liberalization, precarity, and popular struggles.

The rise of the BRICS countries – Brazil, Russia, India, China, and South Africa – has called into question the future of Western dominance in world markets and geopolitics. However, the developmental trajectories of the BRICS countries are shot through with socio-economic fault lines that relegate large numbers of people to the margins of current growth processes, where life is characterized by multiple and overlapping vulnerabilities. These socio-economic fault lines have, in turn, given rise to political convulsions across the BRICS countries, ranging from single-issue protests to sustained social movements oriented towards structural transformation. The contributions in this book focus on the ways in and extent to which these trajectories generate distinct forms and patterns of mobilization and resistance, and conversely, how popular struggles impact on and shape these trajectories. The book unearths the economic, social, and political contradictions that tend to disappear from view in mainstream narratives of the BRICS countries as rising powers in the world-system.

The chapters in this book were originally published as a special issue of *Globalizations*.

Alf Gunvald Nilsen is Professor of Sociology at the University of Pretoria. His research focuses on the political economy of democracy and development in the Global South. His most recent books are *Adivasis and the State: Subalternity and Citizenship in India's Bhil Heartland* (2018) and *Indian Democracy: Origins, Trajectories, Contestations* (2019).

Karl von Holdt is Professor in the Society, Work and Politics Institute, University of the Witwatersrand. Publications include *Transition from Below: Forging Trade Unionism and Workplace Change in South Africa*; *Conversations with Bourdieu: The Johannesburg Moment* (with Michael Burawoy); and *Beyond the Apartheid Workplace: Studies in Transition*, co-edited with Edward Webster, as well as numerous articles. His research interests centre on movements, democracy, corruption, and violence.

Rethinking Globalizations

Edited by Barry K. Gills, University of Helsinki, Finland, and
Kevin Gray, University of Sussex, UK.

This series is designed to break new ground in the literature on globalization and its academic and popular understanding. Rather than perpetuating or simply reacting to the economic understanding of globalization, this series seeks to capture the term and broaden its meaning to encompass a wide range of issues and disciplines and convey a sense of alternative possibilities for the future.

For more information about this series, please visit:
www.routledge.com/Rethinking-Globalizations/book-series/RG

Rising Powers, People Rising

Neoliberalization and its Discontents in the BRICS Countries

Edited by
Alf Gunvald Nilsen and Karl von Holdt

Routledge
Taylor & Francis Group

LONDON AND NEW YORK

First published 2021
by Routledge
2 Park Square, Milton Park, Abingdon, Oxon OX14 4RN

and by Routledge
52 Vanderbilt Avenue, New York, NY 10017

Routledge is an imprint of the Taylor & Francis Group, an informa business

British Library Cataloguing in Publication Data
A catalogue record for this book is available from the British Library

ISBN: 978-0-367-75060-2 (hbk)
ISBN: 978-0-367-75064-0 (pbk)
ISBN: 978-1-003-16085-4 (ebk)

Typeset in Minion Pro
by Newgen Publishing UK

Publisher's Note
The publisher accepts responsibility for any inconsistencies that may have arisen during the conversion of this book from journal articles to book chapters, namely the inclusion of journal terminology.

Disclaimer
Every effort has been made to contact copyright holders for their permission to reprint material in this book. The publishers would be grateful to hear from any copyright holder who is not here acknowledged and will undertake to rectify any errors or omissions in future editions of this book.

Contents

Citation Information

The chapters in this book were originally published in *Globalizations*, volume 16, issue 2 (2019). When citing this material, please use the original page numbering for each article, as follows:

Introduction

Rising powers, people rising: neo-liberalization and its discontents in the BRICS countries
Alf Gunvald Nilsen and Karl von Holdt
Globalizations, volume 16, issue 2 (2019), pp. 121–136

Chapter 1

China's precariats
Ching Kwan Lee
Globalizations, volume 16, issue 2 (2019), pp. 137–154

Chapter 2

Social mobilizations and the question of social justice in contemporary Russia
Karine Clément
Globalizations, volume 16, issue 2 (2019), pp. 155–169

Chapter 3

Mapping movement landscapes in South Africa
Karl von Holdt and Prishani Naidoo
Globalizations, volume 16, issue 2 (2019), pp. 170–185

Chapter 4

Uncovering a politics of livelihoods: analysing displacement and contention in contemporary India
Gayatri A. Menon and Aparna Sundar
Globalizations, volume 16, issue 2 (2019), pp. 186–200

Chapter 5

A precarious hegemony: neo-liberalism, social struggles, and the end of Lulismo in Brazil
Ruy Braga and Sean Purdy
Globalizations, volume 16, issue 2 (2019), pp. 201–215

Chapter 6

Neo-development of underdevelopment: Brazil and the political economy of South American integration under the Workers' Party
Fabio Luis Barbosa dos Santos
Globalizations, volume 16, issue 2 (2019), pp. 216–231

For any permission-related enquiries please visit:
www.tandfonline.com/page/help/permissions

Notes on Contributors

Ruy Braga is Head of the Department of Sociology at the University of São Paulo.

Karine Clément is Researcher at CRESPPA-GTM (Paris) and in the Centre for Sociology of Democracy (Helsinki).

Ching Kwan Lee is Professor of Sociology at the University of California, Los Angeles (UCLA), and Research Associate in the Society, Work and Politics Institute, University of the Witwatersrand.

Gayatri A. Menon is Associate Professor at Azim Premji University's School of Development, and Research Associate in the Society, Work and Politics Institute, University of the Witwatersrand.

Prishani Naidoo is Director of the Society, Work and Politics Institute at the University of the Witwatersrand, and a social activist.

Alf Gunvald Nilsen is Professor of Sociology at the University of Pretoria.

Sean Purdy teaches and researches the history of workers' and social movements in the Americas at the University of Sao Paulo.

Fabio Luis Barbosa dos Santos is an economic historian and Professor in the Department of International Relations at the Federal University of São Paulo, and Research Associate in the Society, Work and Politics Institute (SWOP), University of the Witwatersrand.

Aparna Sundar is an instructor in the Asian Institute, University of Toronto. She is also Visiting Faculty in the School of Development, Azim Premji University, Bangalore, and Research Associate in the Society, Work and Politics Institute (SWOP), University of the Witwatersrand.

Karl von Holdt is Professor in the Society, Work and Politics Institute at the University of the Witwatersrand.

Acknowledgements

We would like to thank the National Institute for the Humanities and Social Sciences (NIHSS), South Africa, for its consistent financial support for this project over a number of years. Without this support we would have struggled to put together the volume and the special issue of *Globalizations* on which it is based.

We would like to thank the School of Development at Azim Premji University for hosting our workshop in Bengaluru in August 2017, as well as the University of São Paulo for hosting our workshop in São Paulo in June 2019, and FAPESP for providing a grant for research in Brazil (project 2017/05588-7).

Alf Gunvald Nilsen and Karl von Holdt
October 2020

Rising powers, people rising: neo-liberalization and its discontents in the BRICS countries

Alf Gunvald Nilsen and Karl von Holdt

ABSTRACT
The rise of the BRICS countries – Brazil, Russia, China, India, and South Africa – has called into question the future of Western dominance in world markets and geopolitics. However, the developmental trajectories of the BRICS countries are shot through with socio-economic fault lines that relegate large numbers of people to the margins of current growth processes, where life is characterized by multiple and overlapping vulnerabilities. These socio-economic fault lines have, in turn, given rise to political convulsions across the BRICS countries, ranging from single-issue protests to sustained social movements oriented towards structural transformation. This article presents an innovative theoretical framework for theorizing the emerging political economy of development in the BRICS countries centred on neo-liberalization, precarity, and popular struggles. It discusses the contributions to this special issue in terms of how they illuminate the intersection between neo-liberalization, precarity, and popular struggle in Brazil, Russia, India, China, and South Africa.

The onset of the twenty-first century has witnessed substantial shifts in the vectors of economic and political power that undergird and structure the workings of the world-system. Whereas the unravelling of state-led developmentalism in the Third World and the collapse of communism in Eastern Europe in the late twentieth century initially seemed to signal an 'end of history' that pivoted around American hegemony, developmental shifts in the new millennium have cast doubt on such diagnoses. It is above all the rise of the BRICS countries – Brazil, Russia, China, India, and South Africa – that have called into question the future of Western dominance in world markets and geopolitics (Nayyar, 2016; O'Neill, 2013; Pieterse, 2018). Mainstream narratives of the economic and political ascent of these emerging powers tend to highlight the potential that this process holds for poverty reduction and progress towards higher levels of human development. Thus, the United Nations Development Programme (UNDP, 2013a) recently celebrated 'the rise of the South' – a process spearheaded by China, India, Brazil, and South Africa – as a progressive and hopeful transformation; and Russia has been widely perceived to be regaining significant economic and strategic ground in a post-communist world (Stuermer, 2009).

More critical voices have questioned the assumption that the rise of the BRICS countries points towards a tendency of developmental convergence in the world economy (see, for example, Kiely, 2007, 2008, 2015, 2016; Starrs, 2014). In part, this scepticism is grounded in the persistence of

Northern power and domination in the global economy (see Hickel, 2017a, 2017b). However, an equally important reason for questioning celebratory accounts of the rise of the BRICS countries is the fact that the changing geography of economic and political power in the world-system is closely related to the emergence of a 'new geography of global poverty' (Kanbur & Sumner, 2012) in which more than 70% of the world's poor now live in middle-income countries (see also Sumner, 2012). Despite impressive growth rates, the southern BRICS countries – Brazil, India, China, and South Africa – are home to more than 50% of the world's poor (Ravallion, 2009). In Russia, poverty has been aggravated by the recent recession, with more than 13% of the population – that is, around 19.2 million people – currently living below the poverty threshold (Agence France-Presse, 2016). Persistent poverty is coupled with very deep and, in most cases, widening inequalities. South Africa is a case in point with a Gini coefficient of 0.631, but China and India have also seen rapidly escalating inequalities in recent years (see also Hung, 2016, Chapter 4; Jayadev, Motiram, & Vakulabharanam, 2011; Oxfam, 2017; UNDP, 2013b; World Bank, 2016). Indeed, recent research shows that Indian inequality is at its highest levels since the early 1920s, as 22% of all income currently accrues to the top 1% of earners (Chancel & Piketty, 2017). Brazil is an exception from this trend – its Gini coefficient declined from 0.594 in 2001 to 0.514 in 2014 (see data.worldbank.org) – but remains a deeply unequal country (World Bank, 2014). And in Russia, the top decile of wealth holders controls 77% of all household wealth, a level of inequality that is equal to that of the USA (Credit Suisse, 2017).

What these statistics ultimately testify to is the fact that the developmental trajectories of the BRICS countries are shot through with socio-economic fault lines. As a result, large numbers of people are relegated to the margins of current growth processes, where life is characterized by multiple and intersecting vulnerabilities rooted in a lack of access to secure and decent livelihoods, the absence of basic social protection and essential public services, and often also the exclusion from established political arenas. Moreover, these socio-economic fault lines have given rise to political convulsions across the BRICS countries, ranging from single-issue protests to sustained social movements oriented towards structural transformation (see, for example, Braga, 2017; Chen, 2014; Clément, 2008; Gabowitsch, 2016; Lee, 2007; Menon, 2013; Naidoo, 2015; Ness, 2015; Nielsen & Nilsen, 2016; Saad-Filho & Morais, 2014; Smith & West, 2012; Von Holdt et al., 2011; Von Schnitzler, 2016). This special issue is dedicated to developing an approach and a set of analyses that can decipher how the developmental trajectories of the BRICS countries generate distinct forms and patterns of mobilization and resistance and, conversely, how popular struggles impact on and shape these trajectories. In doing so, we hope to lay the foundation for a critical conceptualization of the political economy of development in the BRICS countries that unearths those economic, social, and political contradictions that tend to disappear from view in mainstream narratives. To achieve this, the analyses that are offered in this special issue are centred on a triad of key concepts: neo-liberalization, precarity, and popular struggles. Before outlining and discussing these, we briefly introduce the articles collected in this special issue.[1]

All of the authors examine popular mobilization and movements in relation to large historical processes and across a variety of case studies, sites or periods in order to identify longer-term trends, shifts, and possibilities. Ching Kwan Lee examines the changing forms of worker precarity and resistance across three eras of modern Chinese history – state socialism, high-growth market reform, and the current shift to slow growth and overcapacity. Russia followed a very different path of transition from communism, and Karine Clément explores changing popular responses, from the period of shock therapy neoliberalism in the 1990s to the period of growing patriotic nationalism under Putin. Gayatri Menon and Aparna Sundar trace changing forms of dispossession and resistance in India through three case studies, the first two in the period of state-led capitalist modernization and the third in the period of neo-liberal globalization. Karl von Holdt and Prishani Naidoo frame their discussion of South African movements with an analysis of the African National Congress (ANC) domination of the movement landscape, and use case studies of four different moments of mobilization to examine continuities, shifts, and new possibilities. Ruy Braga and Sean Purdy draw out the changing dynamics of popular incorporation and demobilization, followed by both popular and middle-class right-wing mobilizations against the Lulista regime of accumulation, to explain the parliamentary coup against the Workers' Party (PT) president in Brazil. Fabio Luis expands the analysis of neo-liberalization and social conflict in Brazil by examining their role in the expansion of Brazilian companies in the Latin American region, accompanied by super-exploitation of workers and the destruction of the environment. Gathering these articles in a special issue allows us to deepen our understanding of neo-liberalization, precarity, and popular struggle, both conceptually and in terms of the political possibilities they produce.

Neo-liberalization

In many scholarly accounts, the developmental trajectories of the BRICS countries have come to be associated with the term 'post–neo-liberalism' (see, for example, Craig & Porter, 2004; Dale, 2012; Grugel & Riggirozzi, 2012; Harris & Scully, 2015; Sandbrook, 2011; Schmalz & Ebenau, 2012). Crucially, the BRICS countries have been seen as drivers of the emergence of a political economy of development in which market-oriented accumulation strategies are increasingly embedded in modes of regulation that provide social protection and redistribution (see Ban & Blyth, 2013; Ghai, 2015; Nölke, 2012). Such claims, however, tend to disregard the strong continuities that exist between the projects of neo-liberal restructuring that shaped developmental trajectories both in the global South and in post-communist Europe in the 1980s and 1990s and the regulatory regimes that are currently crystallizing in the BRICS countries (Clarke, 2007; Katz, 2015; Prashad, 2012).

Consequently, our approach to understanding the political economy of the BRICS countries in this special issue takes a different view, in which we understand recent interventions in the field of social policy in terms of the 'roll-back' and 'roll-out' dynamics that criss-cross particular and contingent processes of neo-liberalization (see Peck, 2010; Peck & Tickell, 2002). Roll-back strategies are most commonly associated with the onset of processes of neo-liberal restructuring, in which the principal aim is to dismantle regulatory institutions and policy regimes associated with a previous form of state-centred development in order to extend and deepen the reach of the market logic. Conversely, roll-out strategies tend to be spawned by the limits and contradictions of deregulation. Therefore, such strategies tend to be oriented towards enmeshing markets in institutional structures that mitigate market failures and the detrimental social consequences of initial processes of restructuring (see Cammack, 2004; Peck & Tickell, 2007). The extent to which there is an emergent political economy of development in the BRICS countries that is shaped by the kind of interventions associated with roll-out strategies of neo-liberalization, we suggest, must be understood in terms of the crucial tension that animates attempts to reconcile the imperatives of accumulation and legitimation.

Offe (1972) of course identified the reconciliation of accumulation and legitimation as an animating dynamic in the workings of the capitalist state. On the one hand, capitalist states have to ensure that the requirements of accumulation are met by implementing an adequate strategy for economic growth and intervening to adjust imbalances and counter stagnation (see also Jessop, 1990, pp. 198–206). But on the other hand, capitalist states must also ensure democratic legitimacy by gaining the consent and support of their citizens for their mode of governance. As Offe (1984) points out, one of the key ways in which legitimation is achieved is through decommodification in the form of access to social protection and public goods via welfare regimes (see also Borchert & Lessenich, 2016).

By situating the interplay between accumulation and legitimation at the centre of our approach, we move away from a totalizing account of neo-liberalism as a uniform project that is always and everywhere the same[2] and towards an understanding of neo-liberalization as 'a variegated form of regulatory restructuring' (Brenner, Peck, & Theodore, 2010, p. 330). Such an understanding acknowledges the existence of a common denominator of neo-liberalism – namely, that it is a project centred on 'extending market-based, commodified social relations' (Brenner et al., 2010, p. 331) through the deregulation of markets, financialization, and privatization. However, at the same time, it takes cognizance of how neo-liberalization is unevenly developed across 'places, territories, and scales' (Brenner et al., 2010, p. 330), in large part as a result of how neo-liberalizing processes are articulated with the institutional and regulatory legacy of previously hegemonic regimes in the context of specific states (see Cahill & Konings, 2017).

The contributions in this special issue centre neo-liberalization in the BRICS countries in different ways and to different degrees. However, what is common across all the articles is an attempt to bring out how the workings of inter-jurisdictional policy transfer and transnational rule regimes are mediated in and through the encounter with pre-existing political economies and state–society relations in the BRICS countries. This entails an orientation towards discerning how the making of market-oriented forms of regulation have been shaped in path-dependent ways by their encounters with the accumulation strategies, state–society relations, modes of governance, and regimes of citizenship that were forged by state-led developmentalism in Brazil and India, apartheid in South Africa, and communism in China and Russia. It is in these encounters that actual constellations of accumulation and legitimation are configured, and it is by studying them that it becomes possible to unravel how pre-existing political economies and state–society relations have been reworked and transformed at the same time as market-oriented restructuring has come to be patterned in contextually specific ways.

The two post-Communist countries in BRICS have approached the problem of the transformation of their sociopolitical orders and their positioning in the globalized economy in very different ways, as Clément and Lee show. While Russia adopted democratic reforms along with 'violent ultraliberal reforms', including privatization of state industry, producing radical atomization, precarity, and social destabilization in the 1990s, China retained a strong authoritarian Communist regime along with a carefully managed opening to global capital to the 1980s, while retaining state and collective enterprises. Thus, as Clément shows, citizens in Russia were subjected very rapidly to the collapse of old certainties and to a new ideology of individual responsibility and self-blame for their failure to adapt to the new market economy. A revival of rhetoric of patriotic nationalism that implies a rebuilding of state–society relations came later, with the presidency of Putin. In contrast, the Chinese regime retained strong state–society relations and paternalist commitments, and has only shifted rhetoric to promote a culture of individual self-responsibility based on 'innovative entrepreneurship' rather than 'a culture of employment' more recently, with the 'new normal' of slower growth and overcapacity, as Lee argues. These different trajectories have had very different implications for state–society relations and popular mobilization.

India and South Africa were both British colonies, but of very different kinds – the former a colony that achieved independence in the late 1940s, the latter a settler colony which achieved independence under white supremacy in 1910, and national liberation with a black majority government only in 1994. The different character of colonialism in the two countries, and the different periods in which they attained liberation and majority rule, have had implications for their responses to neo-liberalization. India has experienced half a century of state-led modernization and development prior to the era of globalization, while South Africa was plunged into neo-liberalism at the same time as it attained liberation. While India has a strongly established and diversified bourgeoisie which has been able to control the pace of neo-liberalization, South Africa is characterized by a strong and globalized white corporate sector, and a black elite whose development has been stunted by settler domination and now the combined domination of white and multinational capital – tensions which have produced the current political crisis in the country. The two articles presented here show how popular mobilization bears the marks of these different histories and processes of neo-liberalization.

Brazil is also a post-colonial society, but with a much longer history of independence than India or South Africa. Here the key political transition is from military dictatorship to democracy in the late 1980s – roughly coterminous with South Africa's transition from apartheid to democracy, and with similar consequences, as the democratic regime has since its birth been characterized by neo-liberalism. In Brazil, the PT was only elected into the presidency a decade later, in contrast to the immediate ascendance of the ANC in South Africa. However, in both cases, political movements that embodied enthusiastic popular aspirations adopted or maintained neo-liberal macroeconomic policies but attempted to retain legitimacy with mildly redistributive policies and the incorporation of movement leaders and activists into state institutions. In Brazil, the contradictions of these developmental paths are producing shifts and realignments among elites, generating the political crisis of the parliamentary coup, as both Braga and Luiz show, and portending a radical intensification of neo-liberal policies unencumbered by concessions to the popular classes.

While each of the five BRICS countries has pursued trajectories of neo-liberalization marked by specific histories and compromises with existing regimes and social forces – including the popular classes – these different trajectories have all produced new forms of precarity in society, though again with distinct features as well as cross-country resonances.

Precarity

Deepening inequalities across the BRICS countries have to be understood as manifestation of how growth processes have come to be associated with tenacious and in some cases escalating unemployment, widespread underemployment, and increasingly insecure employment (see Denning, 2010; Foster & McChesney, 2012; Ness, 2015). The multiple insecurities that this entails – not just related to inadequate wages and poor working conditions, but also limited access to social protection – have in recent years come to be conceptualized in terms of precarity. As is well known, the concept of precarity was brought to the centre of scholarly debates on the contemporary world of work by Guy Standing's (2011) study of the rise of the 'precariat'. Precarity is on the rise, Standing argues, as a result of the neo-liberal drive 'to create a global market economy based on competitiveness and individualism' (2011, 37).

As much as Standing's concept has set the terms of debate on the nature of work under neo-liberalism, critiques articulated from the point of view of the global South have raised important questions about how precarity is conceptualized. Precarious work, this critique argues, is nothing new in the global South: indeed, it is not tenable in this context to craft an analysis on the basis of a contrast between a precarious present and 'a non-precarious past' (Scully, 2016a, p. 161) in which stable employment, high wages, and access to welfare provisions prevailed (see also Neilson & Rossiter, 2008; Paret, 2016). As Breman and van der Linden (2014, p. 926) point out, casualized and part-time jobs, low and stagnant wages, outsourcing and subcontracting, the substitution of waged work by self-employment, restricted access to welfare, and poorly regulated working conditions have always constituted 'the dominant mode of employment in the developing world' – especially in the informal sector, which is of tremendous importance in many Southern economies (see also Mosoetsa, Stillerman, & Tilly, 2016; Munck, 2013).

This critique is, of course, relevant to the project that we are attempting to articulate in the current issue, where the five countries that are subjected to analysis belong to the global South,[3] and where the emergence of precarious working classes have arguably been absolutely central to recent growth processes (Ness, 2015). We, therefore, approach precarity not as something new and unprecedented, but rather as a constant aspect of work under capitalism, which is constituted in particular ways and present to different extents in particular historical conjunctures and geographical spheres in the world-system. The central substantive analytical challenge then becomes that of disinterring what is specific about precarity in the new political economy that is emerging in the BRICS countries today.

In order to move in this direction, the contributions to this special issue are informed by a two-pronged view of precarity. Firstly, precarity is conceptualized as a material reality. Poverty and inequality, in other words, are seen as being directly linked to unemployment, underemployment, and insecure employment, as well as the extent to which precarious workers have access to public goods and social protection that can mitigate material deprivation. The production of precarity as a material reality is, in turn, understood as a consequence of how specific groups have come to be adversely incorporated into economic and political power structures across spatial scales – from the global, via the national, to the local – as a result of the dynamics of specific processes of neo-liberalization (see Hickey & du Toit, 2013; Mosse, 2010). Secondly, precarity is conceptualized in terms of how it is 'located in the microspaces of everyday life' as 'vulnerability relative to contingency and the inability to predict' the form and direction of life courses (Ettlinger, 2007, p. 320). This entails exploring how precarity is experienced as an inability to rely on work-based incomes to sustain what is considered to be dignified livelihoods and lifeworlds, underpinned by durable social relations in specific places and sites. Crucially, our approach emphasizes how people respond to this experience in ways that are shaped by the intersections of gender, race, caste, and region.

Extant work on this dimension of precarity revolves around mapping how precarity engenders senses of loss, danger, anxiety, and disruption (Han, 2012; Millar, 2014; Neilson, 2015). These are of course very real aspects of precarity as a subjective experience of being in the world. However, abjection does not exclusively define the experiential dimension of precarity. Indeed, life among the urban poor in the global South is characterized to a significant extent by 'quiet encroachments' – that is, the 'silent, protracted but pervasive advancement of the ordinary people on the propertied and powerful, in order to survive and improve their lives' (Bayat, 2000, p. 545). And the significance of oppositional agency in various forms takes us to the last component of our triad of concepts, namely popular struggles, which we discuss further in the next section of this article.

Several of our authors work with fresh definitions of precarity that broaden and deepen the concept. Ching Kwan Lee develops a 'relational and relative concept of precarity' as a condition produced by struggle between workers, employers, and the state, rather than a condition to be identified by a specific set of characteristics. Thus precarity varies over time and in different places. She also extends precarity beyond the traditional focus on the regulation of production to include the politics of recognition, that is to say symbolic or classification struggles over status and the legitimacy of claims, as well as the politics of social reproduction, thus including struggles over the dispossession of land or other means of social security and subsistence. With this concept she is then able to demonstrate that precarity is not a new condition introduced by neo-liberal capitalism, but rather has been continuously produced in different forms and in different ways, differentially affecting various categories of workers, by the state and employers through the three periods of modern Chinese political economy – state socialism, market reform, and the low growth and overcapacity of the contemporary period. In each of these periods the terrain of struggle has shifted: from a struggle over recognition under state socialism, when precarious categories of workers struggled for recognition as full proletarians with all the rights that this entailed; to struggles over the regulation of labour in the period of marketization and integration into the global capitalist economy; and, finally, the emergence of struggles over social reproduction in the current period characterized by increasing unemployment, dispossession, and precarity of livelihoods.

Menon and Sundar undertake a similar move, emphasizing that precarity is a much broader concept with multiple layers of meaning than allowed by the Northern concept of precarious labour. Examining it through the lens of a society characterized by very high levels of informal labour, as well as the attachment of producers to place and the rights of place – land, access to natural resources, an urban pavement to live on and trade from, a stake in an investment fund which provides for generational reproduction – they argue that what is at stake in struggles over precarity is livelihood, a concept that includes the rights to place through which livelihood may be stabilized. Such a definition points to precarity as including multiple processes of dispossession that afflict the labouring classes and petty commodity producers. Like Lee, Menon and Sundar demonstrate that such dispossession has a long history, preceding the advent of neo-liberal capitalism. Through the concept of precarity they explore three different struggles: of fishing villages, to retain control over fishing rights in the face of mechanized trawling; of pavement dwellers, against eviction from places of living and trading; and of garment workers, against the employer efforts to prevent them from accessing their accumulated savings in provident funds.

Clément deepens the concept of precarity to explore its impact on subjectivity in Russia, developing a concept of desubjectivation to describe the loss of self and agency produced by the extraordinary destruction of industry, jobs, and incomes that attended the market transition, as well as the collapse of the institutions and ideological reference points of communism. As Clément puts it, 'people lost the ground under their feet' and had trouble making any sense of the society they lived in or of their place in it. Clément argues that Putin's turn to patriotic nationalism, particularly after the annexation of the Crimea and the confrontations with the West, provided the dominated with a new language and set of reference points through which to 'recover the ground beneath their feet' and understand their society and their place in it, thus in a sense reducing precarity despite the continuity of neo-liberal policies. Clément thus signals the importance of identity and a sense of social place for the possibility of agency, thus returning us to Lee's insistence on the importance of symbolic and classification struggles to the experience of precarity.

The articles on South Africa and Brazil do not directly discuss or redefine the concept of precarity, as do the articles referred to above. However, they implicitly make use of an expanded concept that includes public goods and struggles over social reproduction, such as access to housing, urban land, clean water, and electricity. Both Braga and Purdy's article on Brazil and Von Holdt and Naidoo's on South Africa expand the experience of precarity beyond the marginalized and labouring classes, to include middle-class experiences of a decline in living standards (Brazil) and the financial and debt burden of seeking to enter the middle-class through university education (South Africa).

Braga and Purdy reminds us of Ching Kwan Lee's relational definition of precarity by showing that neo-liberalization is not necessarily directly accompanied by the spread of precarity. Under the 'Lulista regime' of regulation, workers and the urban poor experienced expanded employment opportunities, increased wages and expanding social security through the Bolsa-Família programme, leading to modest redistribution and a decreasing inequality. However, many public goods did not improve in quality or accessibility, and the international financial crisis and the end of the commodity supercycle reversed many of these gains. South Africa exhibits similarly complex dynamics of precarity, with deindustrialization, subcontracting, and outsourcing undermining the conditions of workers and generating large-scale under- and unemployment, at the same time as social policies have focused on large-scale free housing programmes and a huge expansion of social grants.

Taken together, these articles deepen and expand our understanding of precarity in the global South to include dimensions not usually considered in the Northern literature, and also delink the concept from neo-liberalization to include earlier historical periods and social systems not normally considered to be drivers of precarity – colonialism, state socialism, state-led modernization – identifying significant continuities between past and present in place of the rather simple narrative of a rupture with the 'golden era' of full employment and welfare capitalism.

Popular struggles

The contributions in this special issue focus on how precarious workers and poor urban communities constitute themselves as collective actors to contest precarity, and attempt to decipher how and the extent to which their organizing and mobilizing shape the new political economy that is crystallizing in the BRICS countries.

Going beyond the more traditional interest in constituencies, collective identity, and strategic goals, we develop an approach that maps and analyses both broader movement landscapes and the internal life of social movements. The former axis of investigation is intended to direct attention to the way in which specific social movements are embedded in wider 'movement landscapes' (Cox, 2016) and how this embeddedness shapes the form and trajectory of oppositional collective action. This concept is meant to highlight how social movements cannot be understood in isolation, but have to be conceptualized in terms of 'a system of characteristic alliances and oppositions' (Cox, 2016, p. 114) that endure over time and define the context in which both movements and their opponents operate. The second axis directs attention to the micro-dynamics that animate the collective articulation of grievances and claims in public spheres (see Cox & Nilsen, 2014). This entails detailing how objectives, strategies, and tactics emerge through processes of dialogue, debate, and dissent among activists, how forms of leadership and internal hierarchies impact on movement dynamics and, finally, what forms of political skill, knowledge, consciousness, and subjectivity movement participation fosters among precarious workers and poor urban communities.

Bringing these two axes of investigation together and drawing on a quintessentially Gramscian conception of hegemony, the analyses that are developed in the contributions to this special issue seek to determine how popular struggles engage and appropriate hegemonic political institutions and idioms in order to pursue grievances, stake claims, and articulate rights. Simultaneously, we consider how responses from dominant groups and state authorities fuse accommodation through policy changes that concede to oppositional demands with various forms and degrees of coercion with the objective of reproducing a given hegemonic formation (Nilsen, 2015, 2016; see also Nilsen & Roy, 2015). Examining protests through this prism brings to the fore the ways in which movement processes unfold in fields that both enable and constrain oppositional collective action, while simultaneously illuminating how the reproduction of hegemonic formations has to be constantly negotiated in the face of contention (see Gramsci, 1998, pp. 52–55, 180–182; Green, 2002; Mallon, 1995; Roseberry, 1994).

For example, it becomes possible to ascertain whether policy concessions result in co-optation and demobilization, and therefore entrench existing power relations between different actors and groups, or whether they alter balances of power and add momentum to movement processes. It also becomes possible to determine whether specific movement processes have the potential to give rise to political disruptions or whether hegemony is likely to be reconstituted on different terms, and to detail when and why some forms of protest become subject to coercion. It is precisely by unravelling the workings of these equations in and across particular contexts that the approach that we pursue is able to shed some light on how contemporary popular struggles shape the configurations of accumulation and legitimation that undergird neo-liberalization in the BRICS countries. Moreover, it enables us to assess the impact of popular struggle on the forms of adverse incorporation that yield precarity and to ascertain whether incipient patterns of alternative political economies can be discerned within the claims and demands that are being voiced within the practices of social movements in Brazil, Russia, India, China, and South Africa.

Turning to the empirical material and arguments presented in the six papers of this special issue, we discuss here the resonances and tensions between the BRICS countries across three themes: the trajectories of protest and resistance highlighted by each article; the contested languages and claims mobilized from below in each of the countries, and the process of reciprocal appropriation between movements and authorities, as the former attempt to press the justice and legitimacy of their claims, while the latter attempt to absorb and demobilize them; and the prospects for reinvigorated and expanded resistance.

Ching Kwan Lee presents shifting trajectories of mobilization from below across the three periods of Chinese development outlined above. During the authoritarian state socialist period (1949–1979), the categories of workers excluded from the rights and benefits of regular workers (a minority of the workforce) were acutely aware of the discrimination they faced and inferiority of their position because it contradicted so sharply the official ideology of equality and the leading role of the working class, and were able to mobilize their claims during periods of officially sanctioned mass mobilization of the Hundred Flowers Campaign and the Cultural Revolution. During the period of market opening and high growth, the state turned to legal reform and bureaucratic procedures to manage conflict; workers took up these terms to frame their protests and claims, leading to a complex process of collective action (marches and occasional strikes), mediation, and negotiation (in some cases assisted by a new generation of NGOs and in others undertaking independent action), countered by selective repression by the state. Although the 'volume and persistence of worker activism' increased pressure on the state, workers almost invariably avoided moving beyond 'cellular' activism – localized and centred on an individual workplace – avoiding broad mobilization for fear of repression.

Lee thus illuminates a process through which workers appropriated official ideology and institutions – Communist ideology and official campaigns in the first period, legal and bureaucratic procedures and promises in the second – to legitimate their actions and make claims, while authorities responded by incorporating, negotiating, or repressing workers, and on occasion refining forms with new legislation or promises. With the third shift to a 'new normal' of increasing dispossession, indebtedness, disempowerment, and job loss, combined with intensified repression, Lee detects contradictory trends: on the one hand, sporadic instances of 'more violent, volatile and less institutionally incorporated' clashes between precarious workers and the state and, on the other, a trend towards atomization and acquiescence.

Karine Clément detects similar processes through which Russian workers and precarious communities mobilize by framing their protests with the rhetoric presented by authorities, while authorities appropriate the resentment and resistance from below, incorporating these into a new rhetoric of paternalism. While there are many grassroots and labour protests in Russia, they tend to be 'scattered and mostly small-scale', characterized by 'everyday activism' and showing persistent trends towards the localization of struggles, as in the case of China, rather than national mobilizations for social justice. The one exception was the wave of spontaneous protests in 2005 against Putin's attack on the national social benefits system which began with pensioners, and expanded to include a diversity of groups across 100 towns and nearly 80 regions. The Putin regime responded by partially repealing the reform, adopting a new language of social paternalism and launching a variety of health, housing, and education programmes. The recession that hit Russia in 2014 after the annexation of the Crimea has produced widespread hardship, and the Kremlin has accelerated its turn towards a populist and patriotic discourse, successful in mobilizing big demonstrations in its support and against the West. Clément perceives within this the emergence of a new 'social critical' populism from below, as citizens have appropriated the regime's new discourse and have begun to use this language to critique the current order of things. In Clément's analysis, popular nationalist rhetoric from above produces a new social imaginary of a society characterized by cleavages between the elite and the wealthy and the 'hard-working people' and the 'ordinary folk'. The earlier desubjectivation is reversed with a new subjectivation and sense of agency. Thus, in contrast to Lee's analysis of China, Clément sees in Russia both the ongoing vitality of everyday activism, and the beginnings of a new social critique from below which may provide the foundation for more expansive mobilization that articulates explicit demands for social justice.

While Lee's and Clément's respective prognoses diverge, there is a remarkable similarity in their analysis of the ways in which dominated people in the post-Communist states appropriate the rhetoric of authorities in order to claim justice or rights, while authorities in turn adjust positions in response to mobilization, appropriating demands and refashioning them into new policies, new rhetoric, and new promises.

Von Holdt and Naidoo pursue similar themes in their analysis of popular mobilization in South Africa. They develop the concept of 'movement landscape' as a terrain structured by institutions, organizations, symbolic fields, and discourses and laid down by formative action both from above and below, which both empowers and constrains popular mobilization. Popular forms of engagement 'may reproduce prevailing terms of incorporation, negotiate an alteration to them, or transgress them profoundly' – and it is these dynamics they set out to explore through an analysis of four different cases of popular struggle. Essentially they argue that the movement landscape has been deeply structured by the forms of ANC politics laid down during the liberation struggle, as well as the terms and rights established by the new constitution. In all the cases they examine, mobilizations and movements emerge from within the ANC 'constellation of organizations' and tend to adopt the repertoires of action and symbolic discourses that have been sanctified by the same forces, but in the process tend to push up against the structures and forms of the landscape and, in some cases, breach them in different ways. For example, the struggles and massacre of platinum mine workers at Marikana have triggered a process of fragmentation and realignment within the labour movement, and the emergence of unions and federations outside of the ANC constellation, while the #FeesMustFall student mobilizations have proved a fertile crucible for new ideas and symbolic power outside of the ANC tradition, drawing from previously silenced currents of struggle such as black consciousness, the emergence of black feminism, and a focus on decolonization.

This remains a contradictory process, however, as dominant currents within both the student movement and the new labour formations continue to mobilize the popular rhetoric of the ANC, rich as it is in the themes of struggle and resistance, in articulating grievances and making demands, thus reproducing the practices and symbolic universe characteristic of this constellation of organizations, as do the multitude of community mobilizations that operate almost entirely on the terrain of the ANC. This tendency to operate on the terms of the dominant ideology tends to produce multiple localized struggles rather than large-scale mass movements challenging the ANC and its neo-liberal policy orientation – resonating with the analyses of China and Russia summarized above. Von Holdt and Naidoo make visible as well the ways in which the ANC government actively contests the terms of engagement, absorbing organizational leaderships, and refashioning and appropriating demands in the form of new policies in some cases, and in others working to delegitimize struggles and organizations and deploying repressive strategies – most strikingly in the case of the Marikana strikes and massacre, but also in ambivalent ways against the student movement. The authors thus conclude that the high levels of activism and mobilization visible in South Africa 'demonstrate a complex mix of trends', with some tending to 'conserve and reproduce the existing landscape' while others are beginning to refashion the landscape, 'establishing new organizational nodes and repertoires of struggle'.

The case of Brazil, as presented by Ruy Braga and Sean Purdy, presents a dramatically different dynamic of struggle. Whereas in China, Russia, and South Africa our authors analyse tentative and shifting processes of mobilization in the context of politically, ideologically, and institutionally dominant regimes, in Brazil the domination of the PT presidency since 2003 appears to have been more fragile, weakened by the compromises it had to make with a congress dominated by right-wing parties and an extremely confident capitalist class, and only able to exert what Braga describes as a 'precarious hegemony' over the subaltern classes. The latter was maintained with the 'active consent' of the leadership and bureaucracy of the trade union movement and social movements, which were absorbed into government and state institutions with the aim of implementing a progressive developmentalism, and the 'passive consent' of the mass membership secured through progressive social policies such as substantial increases to the minimum wage, the Bolsa-Família and housing programmes, economic growth, and job creation. As in the case of China, Russia, and South Africa, the Brazilian popular movement had been absorbed into the discourse and practices of the PT.

However, this moderated version of neo-liberalism was only able to succeed for as long as it was propelled by economic growth premised in large parts on the global commodity supercycle. As the international financial crisis worked its way through the global economy and the supercycle faded, the PT government of Dilma Rousseff came under pressure to retreat from these social programmes. The result was growing disaffection from below, culminating in the militant strike wave of 2013/2014 and the extraordinary popular mobilization against fare increases in the 'June days' of 2013. While this resulted in substantial wage concessions as well as the withdrawal of fare increases across many cities, the PT government proved unable to absorb or appropriate popular discontent as the regimes of the other three governments discussed above have been able to do. The paralysis of the Brazilian government in face of pressures from the capitalist classes above and popular pressure from below exposed the fragility of PT domination, and culminated in a parliamentary coup by right-wing political parties. Thus, unlike the previous cases, the PT government proved unable to absorb or contain popular pressure beyond the two terms of Lula's presidency, which – in contrast to the previous cases – led to massive explosions of popular protest, including large-scale right-wing mobilizations of middle-class constituencies against the PT. In this context, the capitalist classes and elites felt confident enough to engineer a coup on very flimsy pretexts, and install an aggressively neo-liberal regime which is not only unravelling PT reforms, but attempting to entrench constraints for decades to come. The corruption scandals – which implicate all political parties – and the blatant hypocrisy and self-interest of the political elite have done much to discredit the Brazilian political system. It is not clear in this very new situation what form future popular resistance may take.

Fabio Luis fleshes out Braga and Purdy's analysis of the PT and popular mobilization in Brazil, with an account of its role in the drive to regional integration. Though his project focuses on the expansion of Brazilian business and investment into the region rather than the forms of popular mobilization, he does demonstrate that Brazilian business diplomacy has worked to blunt the radical edge of the Latin American 'progressive wave' at government level, while the environmental devastation, land destruction, and harsh working conditions of Brazilian-funded megaprojects have provoked widespread militant opposition in several Latin American countries, perhaps in some ways mirroring the processes of leadership incorporation and mass protest evident within Brazil.

Gayatri Menon and Aparna Sundar take their analysis of popular mobilization in India in a different direction to that pursued in articles summarized above. Rather than attempt to provide an overview of the development of labour and popular mobilization and its shifting relations with political regimes, they zoom in on three moments of struggle by precarious groups that enable them to explore their specific notion of precarity centred on the concept of livelihoods and place, rather than on labour and the workplace (see above). Each of these moments illustrates what they call a 'regime of dispossession' (following Levien, 2013) – and here they demonstrate that such regimes of dispossession certainly did not make their first appearance with the neo-liberal turn in India, but that they long predated that (no doubt into the colonial period as well). The dispossession of their fishing grounds experienced by the fishers in Tamil Nadu in the form of plunder by mechanized trawlers represented an enclosure of the village commons defined by local custom and notions of village sovereignty – which were in turn mobilized in resistance as the basis for ostracizing trawler captains and merchants, making claims on the state which resulted in legislative changes, and participating in broader alliances against displacement. The second moment, the eviction in Mumbai of some 100,000 pavement traders and dwellers who had previously been displaced from their agricultural land, gave rise to constitutional litigation arguing that the right to life (constitutionally entrenched) included the right of making a livelihood which, in turn, included the right to a place. While the court accepted this expansion of the definition of the right to life, it also permitted the eviction to stand.

These first two moments had their origins in the state-centred national development phase in post-colonial India. The third moment discussed by Menon and Sundar is that of a flash strike of 120,000 mostly female garment workers in Bengaluru in 2016. The strikers avoided trade union involvement; nor were they striking over the extremely low wages and poor working conditions in their industry. Rather, they were striking against a new rule that would have restricted their ability to withdraw savings from their industry provident fund in order to finance their children's education or weddings. Thus, the authors argue, this was not a strike over workplace benefits so much as an entitlement connected to social reproduction, and thus similar to the struggles over livelihood in the first two moments.

Menon and Sundar reject prevailing analyses that typecast these kinds of struggle as petty bourgeois, amorphous and politically unreliable, or alternatively, citizenship struggles, and argue that they constitute 'incipient or emergent' 'forms and languages of contestation' over precarity that suggests a new politics of livelihoods. Finally, they make the important methodological point that researchers such as ourselves should pay attention to the dynamics and meanings of *what is* in the field of popular resistance, rather than imposing a grid of preconceived conceptions about what constitutes a genuine movement or moment of resistance and then seeking evidence for such a development. This may, to a greater or lesser extent, serve as a kind of manifesto for the work presented in this special issue, as well as to the urgent work of charting and analysing the contours and dynamics of the new political economy of development that seems to be crystallizing in the world-system.

Notes

1. The authors whose contributions are featured in this special issue met for the first time in Bergen in 2016 and collectively developed the orientation outlined in this introduction.
2. Examples of such accounts would arguably be Gill (2003), Harvey (2005), McNally (2010), and Panitch and Gindin (2012).
3. We use the term 'global South' loosely so as to include Russia. Clearly reconfigurations of the world-system render the homogeneity of the 'global South' moot.

Disclosure statement

No potential conflict of interest was reported by the authors.

References

Agence France-Presse. (2016, March 22). Millions more Russians living in poverty as economic crisis bites. *The Guardian*. Retrieved from https://www.theguardian.com/world/2016/mar/22/millions-more-russians-living-in-poverty-as-economic-crisis-bites

Ban, C., & Blyth, M. (2013). The BRICs and the Washington consensus: An introduction. *Review of International Political Economy, 20*(2), 241–255.

Bayat, A. (2000). From 'dangerous classes' to 'quiet rebels': Politics of the urban subaltern in the global south. *International Sociology, 15*(3), 533–557.

Borchert, J., & Lessenich, S. (2016). *Claus Offe and the critical theory of the capitalist state*. London: Routledge.

Braga, R. (2017). *The politics of the precariat: From populism to Lulista hegemony*. Leiden: Brill Publications.

Breman, J., & van der Linden, M. (2014). Informalizing the economy: The return of the social question at a global level. *Development and Change, 45*, 920–940.

Brenner, N., Peck, J., & Theodore, N. (2010). After neoliberalization? *Globalizations, 7*(3), 327–345.

Cahill, D., & Konings, M. (2017). *Neoliberalism*. Cambridge: Polity Press.

Cammack, P. (2004). What the World Bank means by poverty reduction, and why it matters. *New Political Economy, 9*(2), 189–211.

Chancel, L., & Piketty, T. (2017). *Indian income inequality, 1922-2014: From British Raj to billionaire Raj?* (Wealth & Income Database World working paper series No. 2017/11). Retrieved from http://wid.world/wp-content/uploads/2017/12/ChancelPiketty2017WIDworld.pdf

Chen, X. (2014). *Social protest and contentious authoritarianism in China*. Cambridge: Cambridge University Press.

Clarke, S. (2007). *The development of capitalism in Russia*. Abingdon: Routledge.

Clément, K. (2008). New social movements in Russia: A challenge to the dominant model of power relationships? *Journal of Communist Studies and Transition Politics, 24*(1), 68–89.

Cox, L. (2016). The southern question and the Irish question: A social movement landscape with migrants. In O. G. Augustin & M. B. Jørgensen (Eds.), *Solidarity without borders: Gramscian perspectives on migration and civil society alliances* (pp. 113–131). London: Pluto Press.

Cox, L., & Nilsen, A. G. (2014). *We make our own history: Marxism and social movements in the twilight of neoliberalism*. London: Pluto Press.

Craig, D., & Porter, D. (2004). The third way and the third world: Poverty reduction and social inclusion strategies in the rise of 'inclusive' liberalism. *Review of International Political Economy, 12*(2), 226–263.

Credit Suisse Research Institute. (2017, November). *Global Wealth Report 2017*. Zurich: Credit Suisse. Retrieved from publications.credit-suisse.com/tasks/render/file/index.cfm?fileid=4B23E1E9-95A7-AA12-2B4C2E0FDFF97EC5

Dale, G. (2012). Double movements and pendular forces: Polanyian perspectives on the neoliberal age. *Current Sociology, 60*(1), 3–27.

Denning, M. (2010). Wageless life. *New Left Review, 66*, 79–97.

Ettlinger, N. (2007). Precarity unbound. *Alternatives, 32*(3), 319–340.

Foster, J. B., & McChesney, R. W. (2012). *The endless crisis: How monopoly-finance capital produces stagnation and upheaval from the USA to China*. New York, NY: Monthly Review Press.

Gabowitsch, M. (2016). *Protest in Putin's Russia*. Cambridge: Polity Press.

Ghai, M. (2015). A model for universal social security coverage: The experience of the BRICS countries. *International Social Security Review, 68*(3), 99–118.

Gill, S. (2003). *Power and resistance in the new world order*. London: Palgrave.

Gramsci, A. (1998). *Selections from the prison notebooks of Antonio Gramsci*. (Q. Hoare & G. N. Smith, Eds.). London: Lawrence and Wishart.

Green, M. (2002). Gramsci cannot speak: Presentations and interpretations of Gramsci's concept of the subaltern. *Rethinking Marxism, 14*(3), 1–24.

Grugel, J., & Riggirozzi, P. (2012). Post-neoliberalism in Latin America: Rebuilding and reclaiming the state after crisis. *Development and Change, 43*, 1–21.

Han, C. (2012). *Life in debt: Times of care and violence in neoliberal Chile*. Berkeley: University of California Press.

Harris, K., & Scully, B. (2015). A hidden counter-movement? Precarity, politics, and social protection before and beyond the neoliberal era. *Theory and Society, 44&45*(5), 415–444.

Harvey, D. (2005). *A brief history of neoliberalism*. Oxford: Oxford University Press.

Hickel, J. (2017a). Is global inequality getting better or worse? A critique of the World Bank's convergence narrative. *Third World Quarterly, 38*(10), 2208–2222.

Hickel, J. (2017b). *The divide: A brief guide to global inequality and its solutions*. London: William Heinemann.

Hickey, S., & du Toit, A. (2013). Adverse incorporation, social exclusion, and chronic poverty. In A. Shepherd & J. Brunt (Eds.), *Chronic poverty: Concepts, causes and policy* (pp. 134–160). London: Palgrave.

Hung, H. (2016). *The China boom: Why China will not rule the world*. New York, NY: Columbia University Press.

Jayadev, A., Motiram, S., & Vakulabharanam, V. (2011). Patterns of wealth disparities in India: 1991–2002. In S. Ruparelia, S. Reddy, J. Harriss, & S. Corbridge (Eds.), *Understanding India's new political economy: A great transformation?* (pp. 81–100). London: Routledge.

Jessop, B. (1990). *State theory: Putting the capitalist state in its place*. Cambridge: Polity Press.

Kanbur, R., & Sumner, A. (2012). Poor countries or poor people? Development assistance and the new geography of global poverty. *Journal of International Development, 24*(6), 686–695.

Katz, C. (2015). Capitalist mutations in emerging, intermediate and peripheral neoliberalism. In P. Bond & A. Garcia (Eds.), *BRICS: An anti-capitalist critique* (pp. 70–96). London: Pluto Press.

Kiely, R. (2007). Poverty reduction through liberalisation? Neoliberalism and the myth of global convergence. *Review of International Studies, 33*(3), 415–434.

Kiely, R. (2008). 'Poverty's fall'/China's rise: Global convergence or new forms of uneven development? *Journal of Contemporary Asia, 38*(3), 353–372.

Kiely, R. (2015). *The BRICs, US 'decline' and global transformations*. London: Palgrave.

Kiely, R. (Ed.). (2016). *The rise and fall of emerging powers: Globalisation, US power and the global north-south divide*. Basingstoke: Palgrave Macmillan.

Lee, C. K. (2007). *Against the law: Labor protests in China's rustbelt and sunbelt*. Berkeley: University of California Press.

Levien, M. (2013). Regimes of dispossession: From steel towns to special economic zones. *Development and Change, 44*(2), 381–407.

Mallon, F. E. (1995). *Peasant and nation: The making of postcolonial Mexico and Peru*. Berkeley: University of California Press.

McNally, D. (2010). *Global slump: The economics and politics of crisis and resistance*. Oakland: PM Press.

Menon, G. A. (2013). Citizens and 'squatters': The contested subject of public policy in neoliberal Mumbai. *Ethics and Social Welfare, 7*(2), 155–169.

Millar, K. (2014). The precarious present: Wageless labor and disrupted life in Rio de Janeiro, Brazil. *Cultural Anthropology, 29*(1), 32–53.

Mosoetsa, S., Stillerman, J., & Tilly, C. (2016). Precarious labor, south and north: An introduction. *International Labor and Working-Class History, 89*, 5–19.

Mosse, D. (2010). A relational approach to durable poverty, inequality and power. *Journal of Development Studies, 46*(7), 1156–1178.

Munck, R. (2013). The precariat: A view from the south. *Third World Quarterly, 34*(5), 747–762.

Naidoo, P. (2015). Between old and new: Struggles in contemporary South Africa. *South Atlantic Quarterly, 114* (2), 436–445.

Nayyar, D. (2016). BRICS, developing countries and global governance. *Third World Quarterly, 37*(4), 575–591.

Neilson, B., & Rossiter, N. (2008). Precarity as a political concept, or, fordism as exception precarity as a political concept, or, fordism as exception. *Theory, Culture and Society, 25*(7-8), 51–72.

Neilson, D. (2015). Class, precarity, and anxiety under neoliberal global capitalism: From denial to resistance. *Theory and Psychology, 25*(2), 184–201.

Ness, I. (2015). *Southern insurgency: The coming of the global working class*. London: Pluto Press.

Nielsen, K. B., & Nilsen, A. G. (Eds.). (2016). *Social movements and the state in India: Deepening democracy?* London: Palgrave.

Nilsen, A. G. (2015). For a historical sociology of state–society relations in the study of subaltern politics. In A. G. Nilsen & S. Roy (Eds.), *New subaltern politics: Reconceptualizing hegemony and resistance in contemporary India* (pp. 31–53). Delhi: Oxford University Press.

Nilsen, A. G. (2016). Power, resistance and development in the global south: Notes towards a critical research agenda. *International Journal of Politics, Culture, and Society, 29*(3), 269–287.

Nilsen, A. G., & Roy, S. (2015). Reconceptualizing hegemony and resistance in contemporary India. In A. G. Nilsen & S. Roy (Eds.), *New subaltern politics: Reconceptualizing hegemony and resistance in contemporary India* (pp. 1–27). Delhi: Oxford University Press.

Nölke, A. (2012). The rise of the 'B(R)IC variety of capitalism' – towards a new phrase of organized capitalism? In H. Overbeek & B. van Aapeldoorn (Eds.), *Neoliberalism in crisis* (pp. 117–137). London: Palgrave.

Offe, C. (1972). Advanced capitalism and the welfare state. *Advanced Capitalism and the Welfare State, 2*(4), 479–488.

Offe, C. (1984). *Contradictions of the welfare state*. (J. Keane, Ed.). Cambridge: MIT Press.

O'Neill, J. (2013). *The growth map: Economic opportunity in the BRICs and beyond*. London: Portfolio Penguin.

Oxfam. (2017). *An economy for the 99%*. Oxfam. Retrieved from https://www.oxfam.org/sites/www.oxfam.org/files/file_attachments/bp-economy-for-99-percent-160117-en.pdf

Panitch, L., & Gindin, S. (2012). *The making of global capitalism: The political economy of American empire*. London: Verso.

Paret, M. (2016). Towards a precarity agenda. *Global Labour Journal, 7*(2), 111–122.

Peck, J. (2010). *Constructions of neoliberalism reason*. Oxford: Oxford University Press.

Peck, J., & Tickell, A. (2002). Neoliberalizing space. *Antipode, 34*(3), 380–404.

Peck, J., & Tickell, A. (2007). Conceptualizing neoliberalism, thinking Thatcherism. In H. Leitner, J. Peck, & E. S. Sheppard (Eds.), *Contesting neoliberalism: Urban frontiers* (pp. 26–50). New York: Guildford Press.

Pieterse, J. N. (2018). *Multipolar globalization: Emerging economies and development*. Abingdon: Routledge.

Prashad, V. (2012). *The poorer nations: A possible history of the global south*. London: Verso.

Ravallion, M. (2009). *A comparative perspective on poverty reduction in Brazil, China and India* (World Bank Policy Research Paper No. 5080). Retrieved from https://papers.ssrn.com/sol3/papers.cfm?abstract_id=1492560

Roseberry, W. (1994). Hegemony and the languages of contention. In J. M. Gilbert & D. Nugent (Eds.), *Everyday forms of state formation: Revolution and the negotiation of rule in modern Mexico* (pp. 355–366). Durham: Duke University Press.

Saad-Filho, A., & Morais, L. (2014). Mass protests: Brazilian spring or Brazilian malaise? *Socialist Register, 2014*, 227–246.

Sandbrook, R. (2011). Polanyi and post-neoliberalism in the global south: Dilemmas of re-embedding the economy. *New Political Economy, 16*(4), 415–443.

Schmalz, S., & Ebenau, M. (2012). After neoliberalism? Brazil, India, and China in the global economic crisis. *Globalizations, 9*(4), 487–501.

Scully, B. (2016a). Precarity north and south: A southern critique of Guy standing. *Global Labour Journal, 7*(2), 160–173.

Smith, J., & West, D. (2012). *Social movements in the world-system: The politics of crisis and transformation.* New York, NY: Russell Sage Foundation.

Standing, G. (2011). *The precariat: The new dangerous class.* London: Bloomsbury Academic.

Starrs, S. (2014). The chimera of global convergence. *New Left Review, 87,* 81–96.

Stuermer, M. (2009). *Putin and the rise of Russia.* London: Weidenfeld and Nicolson.

Sumner, A. (2012). Where do the world's poor live? A new update. *IDS Working Papers, 2012*(393), 1–27.

United Nations Development Programme. (2013a). *Human development report 2013: The rise of the south.* New York, NY: Author.

United Nations Development Programme. (2013b). *Humanity divided: Confronting inequality in developing countries.* New York, NY: Author.

Von Holdt, K., Langa, M., Molapo, S., Mogapi, N., Ngubeni, K., Dlamini, J., & Kirsten, A. (2011). *The smoke that calls: Insurgent citizenship, collective violence and the struggle for a place in the new South Africa; eight case studies of community protest and xenophobic violence.* Johannesburg: Centre for the Study of Violence and Reconciliation & Society, Work and Development Institute. Retrieved from https://www.csvr.org.za/docs/thesmokethatcalls.pdf

Von Schnitzler, A. (2016). *Democracy's infrastructure: Techno-politics and protest after apartheid.* Princeton, NJ: Princeton University Press.

World Bank. (2014). *Inequality and economic development in Brazil.* World Bank Country Study. (Compiled by C. Velez, R. Paes de Barros, & F. H. G. Ferreira). Washington, DC: Author.

World Bank. (2016). *Poverty and shared prosperity 2016: Taking on inequality.* Washington, DC: Author. doi:10.1596/978-1-4648-0958-3

China's precariats

Ching Kwan Lee

ABSTRACT
This essay offers a stylized account of the trajectory of precarious labour in China over the past seven decades and identifies the various contested terrains constitutive of its politics. I define 'precarity' not as a thing-like phenomenon with fixed attributes but as relational struggles over the recognition, regulation, and reproduction of labour. For each of the three periods of contemporary Chinese development, i.e. the Mao era of state socialism (1949–1979), the high-growth market reform era (1980–2010), and the current era of slow growth and overcapacity (since around 2010), I analyse the political economic drivers of precarity – from state domination to class exploitation and then to exclusion, indebtedness and dispossession – and workers' changing capacity and interest to contest it.

While informalization and precarization are global tendencies, the making and meanings of precarious labour take on national, regional, core, and peripheral colorations. This paper examines precarious labour in China, whose 800 million strong workforce (out of a working-age population of about 1 billion) is the largest in the world. Rather than defining precarity or informality as a thing-like phenomenon with a number of characteristics (see e.g. Guy Standing's [2011] list of forms of labour insecurity), and precarious labour as a particular group of workers under specific terms of employment (such as workers without written contract and social insurance), it may be more productive to conceptualize 'informality' and 'precarity' as 'relational struggles'. This means two things. First, the contents and meanings of informality and precarity cannot be fixed as some objective universally applicable indicators, but are always relational and relative, culture- and context-dependent. What is deemed precarious and informal in the United States could very well be considered secure and formal in India, Africa, or China. Just as class is not a structure but a relationship (Thompson, 1963) that happens and changes over time among workers, employers, and the state, to specify the meaning of precarious labour is to specify the kind of relationships entered into among workers, employers, and the state (and perhaps other social actors). Second, these relationships are always the subjects and outcomes of struggles and ongoing negotiations in response to changing political, economic, and ideological conditions. The question of 'struggle' calls for an analysis of the interests and capacities of workers, capital, and the state, and the processes and institutions that embed and reproduce them. It entails both workers' acquiescence and resistance, consent and critique.

For a long time, labour studies have focused on relational struggles between management and workers at the point of production, thus how the law and state policies regulate and condition their power balance and shape class relations. But once we broaden our scope of inquiry beyond full-time workers in formal employment, we will see that there are two other contested terrains besides the *regulation* of production relations. These are the politics of *recognition*, thus the symbolic or classification struggle over who are workers, their normative status and standing, and the repertoire of moral, ideological, and legal claims available to them; and the politics of *social reproduction*, thus care and subsistence provisions for maintaining and renewing workers' labour power on a daily and generational basis.[1] With these three interlinked terrains as an analytical lens, we can see that some workers are disadvantaged and subordinated in all three, while others have varying degrees of security in some or all of them. Precarity then can be analysed as relational and relative, a changeable and changing outcome of struggles along these three dimensions. And, precarization is a process of becoming excluded from the labour market in various degrees, being dispossessed of land or other means of social security and subsistence and denied the recognition of one's labour as labour.

Through tracing the forms of precarity and processes of precarization in China from the state socialist period through the high-growth market reform era to the current new normal of slow growth and overcapacity, this paper shows that these terms apply not just to the era of neo-liberalization. They also apply to state socialist China before its integration with global capitalism, albeit under different institutional arrangements and political economic imperatives. I argue that over the past seven decades, (1) the most salient terrains for relational struggles has shifted from recognition to regulation and now to the social reproduction of labour; (2) the power relation underpinning workers' precarity has changed from state domination to class exploitation and, increasingly, to exclusion, dispossession, and indebtedness.

1. Precarious vs. permanent proletariats under state socialism

Notwithstanding the communist ideology of equality and protection, state paternalism during the planned economy period was practiced on the principle of exclusivity, not universality, resulting in a hierarchy of inequality and insecurity. The famed 'iron rice bowl' – permanent employment with a cradle-to-grave welfare guarantee – was available to only one-fifth of the Chinese workforce, almost all urbanites (Walder, 1986). The vast majority of the working population, including workers in collective industries and the even larger contingent of farmers, were categorically excluded from state-funded and guaranteed welfare. Instead, these workers depended on revenues of their own collective enterprises or communes for wages and collective benefits which varied widely across work units, villages, and regions. The main driver of precarity for these non-state workers in this period was not the market but the Communist state's strategy of accumulation and domination. Worker resistance was spearheaded by marginalized workers who appropriated the communist ideology of equality and proletarian leadership to demand equal recognition, equal compensation, and equal welfare.

Precarity in the Mao era was crucially buttressed by the state-enforced rural-urban divide and the concomitant unequal citizenship regime. The transfer of surplus from agriculture to industry, from country to city, and from peasants to workers could not have been possible without the *hukuo* (household registration) system that essentially locked rural workers down in their birthplace and the state-imposed 'price scissors' which artificially devalued agricultural labour relative to industrial labour. Not guaranteed or supported by state budget, agricultural collectives (i.e. a three-level system of commune, brigade, and team from 1958 to 1979) were self-sufficient basic units of production and accounting, and depended on self-generated resources to buffer risk and provide basic medical services, primary education, and emergency relief (Naughton, 2007). Rural precariousness was starkly displayed during the Great Leap famine: nearly all of the estimated 18–32.5 millions who starved to death were rural residents (Grada, 2011). Based on data on the differential reduction in grain consumption during the famine, it was clear that the state protected urban residents from starvation. Sociologist Martin Whyte (2010) calls this rigid regime of unequal citizenship a 'socialist caste' system (see also Selden, 1993, Chapters 5 and 6).

In cities, during the first three decades of state socialism in post-revolutionary China, a 'dual' labour system separated permanent workers from marginal and temporary workers, with each of these two categories marked by elaborate internal differentiation in wages, benefits, and political status. The much touted 'proletariat masters' of the Communist nation, who enjoyed permanent employment, full and free medical care, housing and a pension amounting to more than half of their former wages, represented only a small minority of the Chinese workforce at any point in time, and were found only in the urban, state-owned, heavy industry sector. The split and inequality between the regular and the contract proletariat co-existed inconveniently and incongruously with the official ideology aimed at creating a united proletarian political backing for the Communist Party. The contradiction between reality and ideology – between policies geared to incentivize productivity by differential compensation and policies aimed at realizing 'work according to need' and protection for all – surfaced most publicly during mass mobilization of the Hundred Flowers Campaign (1957) and the Cultural Revolution (1966–1976). The 'contract proletariat' was at the forefront of labour activism, seizing these state-endorsed moments of class struggle, to demand equal treatment in wages, benefits, and permanent terms of employment.

Several political economic conditions led to institutionalized inequality among Chinese workers. During the revolution, the Chinese Communist Party drew its working-class support mainly from southern skilled artisans – printers, copper fitters, metal workers, and mechanics – whose guild tradition of exclusivity and paternalism found expression in the new communist industrial order in the People's Republic. Former leaders of the communist labour movement in Shanghai, the industrial heartland of pre-revolutionary China, became top officials in charge of instituting labour insurance regulations and according trade unions with important welfare functions.

> But just as only a portion of labor had been actively engaged on the communist side during the revolution, so the fruits of struggle were enjoyed by a limited constituency as well … In 1952, when the new labor insurance system was first implemented, a mere seven percent of the work force was covered by its generous provisions. By 1958, following the socialization of industry, coverage reached a high point of thirty percent. In 1978, at the beginning of the post-Mao reforms, only some twenty-two percent of the labor force could claim such benefits – a figure that remained steady throughout the 1980s. (Perry, 1996, p. 67)

Contrary to its connotation, the 'planned' economy had to deal with financial constraints, production pressure, input shortages, and fluctuation by creating flexibility in its work force. It also depended on the deliberate use of unequal rewards to incentivize productivity among workers, spawning different kinds of polarities within the labour force, across sectors (light and heavy industries and service), ranks (seniority), occupation (skills), and ownership type (state or collective). On the eve of economic reform, there were 13 million temporary workers (or 16%) in industrial employment alone (Walder, 1986, p. 41). A bewildering number of informal arrangements allocated these urban and rural residents to different kinds of temporary positions to provide necessary flexibility to state industries under the planned economy. *Temporary* workers were needed to do work permanent workers resisted doing, to pitch in during hot summer months when absenteeism of permanent workers was common, to undertake enterprise expansion or building addition, etc. Then there were the *apprentices* who endured years of training at sub-standard wages and benefits and were often resentful of their masters. *Migrant workers* from the countryside took up contract jobs in the cities, receiving salaries without any benefits. Their numbers expanded rapidly during the Great Leap Forward. 'Social youth', a euphemistic term for the unemployed youth, usually of urban bourgeois family background, who refused to go into agriculture, were encouraged to join propaganda work to ensure their political loyalty. City governments set up labour service stations, which functioned as labour contractors and charged service fees, to help people looking for temporary jobs. In Shanghai, China's premier industrial centre, 'the social division between secure and marginal workers is as notable in a developing Communist city as in a developing capitalist one' (White, 1976, p. 115).

If the state socialist strategy of accumulation called for instituting a hierarchy of rural and urban precarity and vulnerability, its legitimating ideology directly contradicted this reality and informed workers' consciousness and claims. Precarious workers under Communism developed heightened consciousness of their class position and disadvantages ironically exactly because official propaganda trumpeted equality and unity. Historians of Chinese labour have established that marginal workers played a disproportionately active role in responding to significant episodes of political mobilizations – the Hundred Flowers, the Cultural Revolution and the April Fifth Movement in 1976. In 1957, after a national outpouring of labour unrest in 1956, partly spurred by popular dissent during the Hungarian revolt, labour disturbance erupted in more 587 enterprises involving nearly 30,000 workers. Workers in 'joint ownership' enterprises, apprentices, temporary workers, and those who had lost their permanent status through job reassignment, all resentful of their inferior conditions of service, drove the unrest (Perry, 1994). A decade later, 'the economistic wind' (workers demand for material improvement) during the Cultural Revolution originated among

> long-term irregular workers and those workers who had been mobilized to go down to the countryside to support the peasants. Later the demands of these groups spread to workers in the interior and to intellectual youths who had been part of the 'up to the mountains, down to the countryside' resettlement campaign. Eventually the economistic fever infested even permanent state employees with secure urban household registration. (Perry, 1996, pp. 72–73)

Then, in 1976, mass demonstrations and riots broke out in more than 40 places across the country. Young and marginalized workers who were persecuted for their bourgeois leanings during the Cultural Revolution seized this occasion of commemorating the late Premier Zhou Enlai to express their dissatisfaction with the political persecutions and injustices they suffered (Heilmann, 1993).

In short, notwithstanding the mythology of communist egalitarianism, worker solidarity, and state paternalism, the Chinese working class under Mao was fragmented and marked by inequalities in the realms of production, social reproduction, and social status. On top of pre-revolutionary cleavages of gender, skills, and native place origins, the Communist party-state sponsored and solidified labour divisions along lines of state or collective ownership, core and peripheral industrial sectors, rural-urban *hukuo*, party and non-party membership, and permanent and temporary status. The state, or its politics and policies, was the main driver of protection and precariousness, both material and symbolic. The centrality of ideological domination in the Mao era and the glaring contradictions between socialist ideology and reality fuelled working-class discontent and resistance. Workers were able to seize the moments when the political opportunity structure was opened up by elite struggles at the top. Relational struggles of precarity in this period pivoted on recognition targeting the state, thus marginalized categories of workers leveraging symbolic resources offered by official ideology to make material claims on the state. Obtaining official recognition as formal and permanent workers was key to improvement in power in the workplace and better welfare. As we shall see, in the reform era, with the overhaul of the permanent employment system and the ideological ascendance of market competition and individual responsibility, workers' recognition struggles gave way to regulation struggles in the workplace as the Communist regime turned to the law to regulate the market economy and employment relations.

2. High-growth market reform era: 1980–2009

If the driver of precariousness and protection of worker livelihood during the Mao era was the state, China's reform and opening since around 1980 has ushered in the global capital as an added force of precarization. To catch up with the developed world and finding its competitive niches in the lowest nodes of the global production chains, China's industries and workers bear the disproportionate costs (i.e. razor-thin profit margins and exploitative labour conditions) of global capital's flexible accumulation. Beyond global industries, Chinese domestic strategies of growth (i.e. fixed asset investment and state-led urbanization) have also led to the rise of precarity in construction and urban services, while its strategy of domination (i.e. by monopolizing representation of worker interests) and of legitimation (i.e. market-driven trickle down developmentalism) have seriously hampered the bargaining power of labour vis-à-vis capital. If in the pre-reform period, state domination via ideology fuelled recognition struggles, in the reform period, labour laws mediate and mitigate class exploitation, and regulation becomes the pivotal contested terrain of precarious labour struggles.

This section on the high-growth phase of Chinese reform (1980–2010) first depicts a spectrum of informal labour modalities in manufacturing, construction, and services, and discusses how their emergence is predicated on the state's economic development strategies. While the Chinese labour literature has spotlighted the archetypal semi-proletarianized migrant worker in global factories, this section brings to light less visible (i.e. less recognized) forms of precarious labour – the self-exploited, 'rush order' micro-entrepreneurs, student interns, dispatch workers, construction workers, street vendors, and care workers, amongst others. The second part discusses the state's strategies of legitimation and its alliance with various types of capital in the making of precarious labour. Contrary to the commonplace understanding of precarious labour as the absence of state regulation, I show that the state is actively involved in the relational struggles that define precarity in China. The third part of the discussion turns to workers' capacity, interests, and activism, a constitutive moment of precarity. As the state used the law and its elaborate bureaucratic apparatus (arbitration, mediation, and petition systems) to regulate class conflicts between capital and labour, legal mobilization also became the prevalent mode of worker struggles. When these channels fail to resolve conflicts, the state would resort to bargaining with protesting workers or selective repression to maintain social stability. The strong performance of the economy gave the state the fiscal capacity required for economic absorption of labour conflicts, as well as shaped workers' interests in opting for 'exit' (i.e. job hopping) as a strategy of survival.

2.1. A spectrum of precarity: factories, ghost workshops, and shadow workers

After the crackdown of the 1989 Tiananmen uprising, the Communist regime simultaneously confronted a legitimacy crisis and a severe economic downturn. In response, the Deng Xiaoping leadership made a decisive move in the early 1990s to hasten the pace and scope of economic liberalization and internationalization. The first casualties of urban reform were state workers in old industrial regions. But the death of the socialist working class also saw the birth of a new working class made up mostly of migrants from the countryside (Lee, 2007). By then, global capital had consolidated a regime of flexible accumulation, spinning commodity chains around the world, with profits reaped mostly by multinationals in advanced core countries that specialized in design, brand, and market development. The logistical and information technology revolutions had made global sourcing and contract manufacturing the paradigmatic organizational mode of capitalist production. China found a niche as 'the workshop of the world', thanks to its large, disciplined and relatively educated and healthy rural work force – legacies of the state socialist period (Arrighi, 2008). There-fore, precarious labour in the reform period resulted partly from the historical timing of China's insertion into the global economy where it found a competitive edge in the lowest nodes of the com-modity chain.

Nike, Gap, Apple, Samsung, Wal-Mart, and the likes stand at the commanding height of many 'buyer-driven commodity chains' that have extensive networks and elaborate hierarchies of contract manufacturers and subcontractors in China. The extreme concentration of profits at the top and the human consequences for achieving production flexibility dictated by the consumer markets, all at razor-thin profit margin at the bottom, was most starkly revealed in investigative reports on the manufacturing of the iPhone in the wake of worker suicides at Foxconn, the China-based contract manufacturer for Apple and other top brands of personal electronics. Breakdowns of the cost struc-ture of an iPhone show that Apple pockets 58% of its sales price, while Chinese labour cost accounts for only 1.8%. Foxconn's profit margin was 1.5% in 2012 when Apple's was 39.3% (Chan, Pun, & Selden, 2013, p. 7). A series of *New York Times* reports recounted how phenomenal flexibility on the Chinese shop floor responded to Apple's CEO Steve Jobs' whimsical injunction 'I want a glass screen', just six weeks before the first-generation iPhone hit the market. Executives at Apple admitted that there was no other place to go except China where an assembly line overhaul happened literally overnight (Duhigg & Bradsher, 2012).

The despotic factory regime that exploits and disciplines tens of millions of Chinese migrant workers has been the focal concern of China labour studies in the past two decades. In reality, hidden within and beneath this factory regime are many other modes of precarious work. For instance, since the mid-2000s, global and domestic factories have increasingly turned to a new vulnerable group of informal workers – student interns. In Foxconn and Honda factories, interns account for 15–50% of the workforce. These are students enrolled in vocational schools' nursing, auto maintenance or business administration programmes, but are sent to these factories for 2 months to 2 years as a man-datory part of their training (Chan, Pun, & Selden, 2015; Zhang, 2015). Working without labour con-tract or social insurance, doing tasks unrelated to their majors, these workers are not recognized as workers under the Labour Law, although they work and live like other full-time workers.[2]

What has also escaped media and scholarly attention are the numerous, layer upon layer of sub-contractors working for global contractors in a wide range of industries. Buffering suppliers of global companies from market fluctuations, and concealed in shadowy workplaces, are many modalities of informal production arrangements, ambiguous class relations, and precarious livelihoods that defy binary categorization of 'labour' and 'capital'. Chinese sociologists Huang Yan, Fan Lulu, and Xue Hong have discovered a hidden world of mobile 'rush-order' workshops (Fan & Xue, 2017; Huang, 2012). Kin, familial, and locality ties and trust, not legal contracts, bind workers together as 'on call' mobile but skilled work groups. They show up in subcontractor factories to complete par-ticular rush orders. Some even show up in factories with their own sewing machines and production equipment they bought from other 'on-call' enterprises that moved on when orders disappeared. Hence the numerous 'factory-for-sale' advertisements plastered on public walls in many industrial areas.

Experienced and well-connected workers have become micro-entrepreneurs toiling alongside family members in rented workshops. Lacking employment security and insurance protection, but working at an intense pace, rush order workers are reported making more money than regular factory employment, if and when orders exist. These workers and the factories that hire them seldom show up in industrial statistics.

> Most small factories like ours are not registered businesses. We do not issue invoices, so officials from Industry and Commerce Office, as well as the Tax Office, rarely come to visit. But the Labor Department does come to inspect regularly, so we do need to offer a bribe on different occasions, otherwise they will just come to check our labor agreements, pension, child labor, and etc. (Huang, 2012, pp. 198–199)

Many of these on-call micro-enterprises or worker cooperatives have emerged as vast networks or clusters of production: garment in Humen, Dongguan; electronics assembly in Shijie, Dongguan; leather in Shiling, Huadu District of Guangzhou; lighting fixture in Gu, Zhongshan; footwear in Wenzhou, Zhejiang; and textile in Shaoxing, Zhejiang. Native-place networks also bring migrant workers from particular hometowns to corner labour market niches: workers from Hubei province's Jingzhou City are engaged in Humen's garment industry; Jiangxi province's Ganzhou in Dongguan's electronics assembly; and Hunan Province's Shaoyang in Huadu's leather industry.

2.1.1. Construction workers

Besides China's niche in the global value chain, the centrality of state investment in infrastructure as a motor of economic growth has contributed to a three-decade long construction boom. Between 1978 and 2008, fixed asset investment grew from 30% to 45% of GDP, whereas household consumption dropped from 50% to 35% (Lin, n.d., p. 9). The $570 million stimulus package Beijing rolled out after the global financial crisis in 2008 created another infrastructure construction binge, in a sector already plagued with overcapacity (Naughton, 2009). In 2010, construction accounted for some 25% of China's GDP. A steady 30–50% of the 260 million strong migrant workforce have found employment in construction which is also the number one industry employing male migrant workers (Swider, 2015b, pp. 4–5).

Worldwide, construction is one of the most informally organized industries, thanks to its project-based, mobile nature, its intricate, labour-intensive work process requiring a plethora of skills, and a long tradition of extensive subcontracting through labour brokers. Three types of informal employment configurations can be differentiated (Swider, 2015b), revealing the slave-like conditions for those workers relegated to the bottom-tier of this hierarchy of informal work. The least vulnerable condition, what Swider calls 'mediated employment', is where the employment relationship is established, mediated, and regulated through a contract labour system based on standardized, widespread, yet informal agreement. Then there are those operating under 'embedded employment' which regulates work and workers through social networks. Finally, under 'individualized employment', workers find employment through street labour markets and face despotic employment relations regulated through violence or the threat of violence. While the Chinese press and the Chinese government have exposed the rampant problem of wage non-payment experienced by the first two types of informal employment in construction, the blatant abuses suffered by the last category of workers have gone under the radar. Most of the time these workers work for food and shelter rather than wages. When they get paid, they are paid at a piece rate that requires an inhumane pace of work and long hours. The main control mechanism is violence and the workers' main alternatives are begging or criminal activities.

2.1.2. Service: street vendors, domestic workers, and dispatch workers

Besides heavy investment in infrastructure, the Chinese state's growth strategy through break-neck urbanization has also generated a sizable informal service economy in its major global and metropolitan cities. At times visible, at times not, subsistence, low-wage or 'wage-less' labour of self-employed petty commodity traders, street vendors, maids, personal service providers of all kinds meet the cities' consumption and entertainment needs (shopping, strolling, and socializing). In Chinese cities, much like Indian and African ones, informalization of formal sector employment reduces workers' security and purchasing power and increases their dependence on the urban commons as a site of work and consumption. In 2010, there were an estimated 18 million street vendors in China's urban areas or 5.2% of the urban workforce, and 16% of those in urban informal employment (Huang, 2015, p. 64). In major cities in the Pearl River Delta in South China, street vending has become a major survival strategy among migrant workers who account for 90% of street vendors. In Guangzhou, for instance, an estimated 1.7 to 2.7 million out of a total of 4.2 million rural migrant workers in 2008 were informally employed (Wan, 2015, p. 17). About 60% were male. Many of them worked as street vendors of food, fruit, and consumer commodities, waste and trash sorters, cooks and servers in small restaurants, hair stylists, porters, motorcyclists, itinerant interior decoration workers, etc.[3] Domestic workers, another prevalent mode of informal work, reached 20 million in 2015, according to government statistics (Dan, 2015; IIEC, 2015). Most of these are middle-aged female migrant workers or laid-off urban workers.

Finally, dispatch workers, or agency workers, emerged only in the late 1990s when the government encouraged 'flexible employment' in response to mass unemployment induced by the restructuring of state-owned enterprises. By 2012, there were an estimated 37 million dispatch workers accounting for 13.1% of registered employees. The trend of increasing prevalence is particularly visible in the service sector (All-China Federation of Trade Unions, 2011). Even though dispatch workers are defined and regulated by the Labour Contract Law of 2008 and their protection augmented in the revised Labour Contract Law of 2013, widespread violation and evasion of the law by employers is well documented. Most ironically, state-owned enterprises are found to be major users of dispatch workers (China Labour Support Network, 2014).

2.2. The state's strategies of domination and legitimation

Thus far, I have argued that global capital's flexible accumulation coupled with the Chinese state's strategies of economic development have given rise to a spectrum of precarious labour conditions in manufacturing, construction, and services. Next, we need to consider the state strategy of subordinating labour by (1) monopolizing workers' representation and repressing workers' autonomous activism; (2) legitimizing worker subordination by an ideology of trickle-down developmentalism; and (3) regulating the supply, rights and entitlements of workers through national labour laws and migration policies, while local governments strategically adapt them to align with capital's variegated interests.

2.2.1. Repressive representation

The All-China Federation of Trade Unions is the only legal workers' union in China. Organized as a complex hierarchy extending from Beijing to each province, municipality, district, and enterprise, with its personnel tightly controlled by the Chinese Communist Party at the national and local levels, and dominated by management at the enterprise level, the ACFTU is deeply alienated from its 285 million strong rank-and-file workers. Union membership typically includes management from whom most of the union chairs at the enterprise level are appointed or indirectly elected. Unions are financed by a 2% payroll tax paid by the enterprise rather than membership dues (Friedman, 2014). Above the enterprise level, union cadres are recruited through the same civil service examination as all other government officials, and they behave and think like government officials. As Eli Friedman describes, 'union officers' first response to strikes is that of an agent of the state: intervene, 'rationally' encourage dialogue, convince the workers to make 'reasonable' demands … and perhaps try to persuade management … to meet some of the workers' demands' (Friedman, 2014, p. 55). According to the Trade Union Law, the union should assist the enterprise to resume production and work order as quickly as possible. This monopolization or appropriation of worker representation by the party-state deprives Chinese workers of a powerful leverage to bargain with capital, buttressing an institutional foundation for precarious labour to spread in China.

In the shadow of the official trade unions, grassroots labour NGOs have proliferated slowly but steadily since the late 1990s. Nationally, there are an estimated 40 or so labour NGOs operating semi-legally in major industrial regions to provide legal counselling, training, and recreation services to migrant workers. Reliant on foreign foundations and domestic donations, established by concerned academics, journalists, or former workers turned rights activists, these NGOs lead a very precarious existence in the legal limbo (Lee, 2017). Harassment and crackdowns by officials and employers, even physical assaults by thugs, are commonplace.

2.2.2. Trickle down developmentalism

For three decades, official ideology has prioritized economic development above equality, justice and other collective goals. Each top leader formulated his own 'theory' which was then summed up in catchy slogans for mass consumption, all of which expressed, one way or another, a trickle-down vision of economic development. From Deng Xiaoping's 'development is the hard truth', 'let some people get rich first', to Jiang Zemin's 'three represents' (the CCP represents China's advanced productive forces) and Hu Jintao's 'total devotion to development', from Beijing to localities, achieving GDP growth is the cardinal task of government. In the process, violation of workers' rights, abrogation of state welfare responsibility, and destruction of cultural heritage have found justification in the belief that development is the solution to social problems. During the high growth period, the general rise in income and living standards indeed lend credence to the idea of trickle-down, even as inequality worsened with China's Gini reaching 0.61 in 2010, putting China at or near the top of the world league tables (Whyte, 2014).

2.2.3. Legal regulation and absorption of labour conflict

A common conceptual error in the literature is that precarious and informal labour is caused or defined by the absence of state regulation. Quite the contrary is true in China and elsewhere, where the state is central to the structuring and reproduction of precarity through laws and government policies. Three examples illustrate the alignment of state and capital interest in legalizing precarious labour but according workers inferior rights and entitlements. First, the Labour Law and the Labour Contract Law stipulate inferior treatments for dispatch workers as opposed to regular workers, by defining the formers' relation to employers as one of 'labour service' relation rather than 'labour' relation. As 'employees' rather than 'workers', dispatch workers are legally excluded from social insurance and other labour protections stipulated in the Labour Law (Pang, 2016). Second, the state actively regulates the supply of the massive migrant population (about 250 million today). Their rights and entitlements are stipulated by local government policies in different regions, forming a variety of citizenship regimes that reflect the need of capital for a particular kind of workers (Wu, 2017). Third, the supply of student interns is orchestrated by the Ministry of Education, vocational school administrators and teachers, local education departments and mediated by private labour agencies (Chan, 2017).

More generally, since China's first ever National Labour Law took effect in 1995, and the subsequent promulgation of a series of specific laws on labour contract, labour dispute mediation and arbitration, employment promotion and anti-discrimination, the legal arena has become the main site of labour struggles. On paper, Chinese labour legislations set such a high standard that according to an OECD report on employment protection, China in 2008 ranked second in employment protection across ten major developing economies and exceeded the OECD average substantially (Gallagher, 2014). The problem is, rather than submitting itself to the rule of law, the Chinese state, both central and local, uses the law as an opportunistic instrument to achieve policy and political goals. This means that labour laws are enforced if they are in the interest of the government at various levels of the political system. A few official statistics illustrate the gaps between legal rights and actualized rights among migrant workers. In 2014, 62% of migrant workers still lacked written contract, 84% lacked pension, 83% health insurance, and 90% unemployment insurance (Li, 2016, p. 69).

2.3. Regulation struggles: workers' legal mobilizations and protest bargaining

The pivotal role of the law in regulating labour market and labour relations channels labour resistance to the terrains of the law and related bureaucratic institutions, as workers pursue legal and policy mobilization. The state uses labour resistance as a 'fire alarm' mechanism that alerts local government to particularistic and particularly egregious labour violations (Gallagher, 2014). When the volume of labour disputes points towards certain serious abuses by employers, the Central Government resorts to another round of legislation requiring more stringent labour protection, triggering new responses by employers to bypass the new legal constraints on their use of labour. In this process, both the state and the employer have a common interest in pre-empting workers from developing collective organization capacity.

Industrial workers have been most prominent in labour unrest. In the 1990s, rustbelt workers took to the street making moral economic claims often enshrined in state regulations about their health care and pension benefits. In the sunbelt, tens of millions of young migrant workers employed in export-oriented factories waged their own struggles against exploitative labour practices and violations of their legal labour rights – non-payment of wages, excessive overtime, unsafe workplaces, arbitrary dismissal, and dehumanizing shop-floor discipline. In both cases, labour unrest – taking the form of street protests, public demonstrations, road blockage, strikes, and legal mobilizations – was characterized by localized, single-factory cellular activism, with workers making legalistic claims as rights bearing citizens, with no escalation in terms of lateral coordination.

On the part of the state, social stability has been maintained by a deft combination of protest bargaining (buying workers off during mobilization), bureaucratic absorption (channelling workers into mediation and court procedures), clientelism (exchanging cooperation for material advantages), co-optation (recruiting workers as party members and sponsoring NGOs under official banners), and selective repression (arrests and harassment of influential activists as warning for all) (Lee & Zhang, 2013). State policies and market development fragment workers' interests and identities, while repression and co-optation have largely been effective in crashing and pre-empting any attempt at cross-enterprise, cross-class, cross-regional, and cross-sectoral mobilization. Over time, even without institutional empowerment, the volume and persistence of worker activism has created pressure on the state to improve workers' lot – from the establishment of minimum wage regulations and state provision of minimum livelihood guarantee, to the promulgation of various labour laws and steady increase in wage levels (Solinger, 2009).

3. Into the void: authoritarianism and slow growth (2010 to the present)

Since the global financial crisis of 2008 and a prolonged downturn in China's traditional export markets, many internal imbalances of the Chinese economic 'miracle' have been exacerbated (Hung, 2015). If sustained economic growth has buttressed the legitimacy of one-party authoritarian rule for three decades, China is certainly entering unchartered waters. Concurrent with what the government has called an *economic* new normal, a *political* new normal – the state's repressive turn against civil society – has also taken roots since 2012 when President Xi Jinping took power. In this new phase of slow growth but augmented authoritarianism, labour has been hit hard. I argue that as more workers fall outside the recognition and regulatory framework of the law, the most salient contested terrain will shift to the social reproduction of labour, or livelihood itself. Beyond exploitation, more workers are compelled into relations of dispossession, indebtedness, and exclusion. These relations may reorganize labour's interest and capacity in new ways, and spur labour activism to take a more disruptive and volatile turn outside institutionalized and regulated arenas. They may also prompt the state to reform its social protection policies in order to pre-empt a livelihood crisis for many in a period of economic downturn.

3.1. The 'new normal'

The Chinese government has officially announced the end of the high-growth period (Bradsher, 2012). The 12th Five Year Plan (2011–2015) recognized that an annual growth in excess of 10% (the average over 2003–2010) was unsustainable and envisaged the annual growth rate to be around 7%, which was further revised down to 6.5% in the 13th Five Year Plan (2016–2020) (USCBC, 2010; APCO Worldwide (2015)). Plagued by overcapacity in steel and coal and other 'zombie' state-owned industries, the government announced in 2015 a scheduled massive lay-off of 5–6 million workers in 2016 (Lim, Miller, & Stanway, 2016). Top officials in Beijing have blamed the Labour Contract Law for creating rigidity and neglecting business interests, while some local governments have frozen wage increases and reduced employers' contributions to social security accounts. The government has signalled its intention to revise the labour law to reduce protection for labour and create more labour market flexibility in the face of economic slowdown (Wong, 2016).

The challenge for the working population is much more complicated than a sheer reduction in aggregate growth rate and less demand for labour. Besides labour market exclusion and precarization, two other kinds of power relations – dispossession and indebtedness – contribute to precarization in social reproduction.

3.1.1. Dispossession

In the wake of the 2008 global financial crisis, the Chinese government rolled out an aggressive stimulus package equivalent to 12.5% of China's 2008 GDP, to the tune of US$586 billion. This unleashed a period of debt-fuelled growth whereby local governments borrowed heavily from state banks to fund transport and power infrastructural projects, build housing, and invest in rural health and education. While these measures stabilized the economy in the short run, they also exacerbated the problems of overcapacity and local debt. Local governments have since relied more heavily on selling land to repay their massive debts and service interest payments, leading to rampant land grabs which were intensified by another state policy to stimulate domestic consumption – state-led urbanization.

The latest push for farmers to get off the land and move to the cities came from the 'National New-type Urbanization Plan', announced in 2014, which aims to elevate China's urbanization rate from 54% to 60% of the population by 2020. The rationale is simple: to boost domestic demand and increase consumption. Even though some economists have criticized this logic as flawed – usually development leads to urbanization, not the other way around – in the calculation of Primier Li Keqiang, 'every rural resident who becomes an urban dweller will increase consumption by more than 10,000 yuan (US$1587) … there remains a massive untapped labour pool in the villages, leaving great potential for domestic demand as a result of urbanization' (Li, 2013).

As a result of land grabs and state-enforced urbanization, a double crisis of land dispossession and unemployment is spreading among farmers who have moved to the cities from the countryside. Land grabs have happened in 43% of the 1791 villages sampled in a multi-year 17-province survey (Landesa, 2012). One recent ethnographic study depicts the grim reality for migrant workers after their land was dispossessed. In Sichuan, one of the largest labour-sending provinces in China, they became the most undesirable workers for labour brokers in the construction business. Since labour brokers have to underwrite the cost of transportation and living during workers' employment period and labourers must survive until the end of the year for wages to be paid, landless workers are seen as too precarious for this precarious occupation. 'Without land, brokers and labourers face new financial pressure. Brokers must shift recruitment to other sites where labourers hold land and are better able to withstand precarious employment' (Chuang, 2015). In short, China's landless migrant workers, now nominally urban residents in townships, find themselves in an emerging underclass position that is even more precarious than that of the conventional landholding migrant worker.

3.1.2. Indebtedness

Just as jobless growth is a global trend, the Chinese government's response conforms to trends elsewhere – promoting entrepreneurship and the gig economy. To manage popular expectation about a prolonged economic slowdown, and to create a culture of entrepreneurship rather than a culture of employment, the Chinese Premier announced in his 2015 Government Work Report that 'innovative entrepreneurship' is the 'new economic normal' for Chinese citizens. From 2014 to 2015, three and a half million new private business entities were formed, 90% of them micro-enterprises in information, software, entertainment, and services (Rungain Think Tank of Entrepreneurship, 2016). Laid-off women workers were enticed to become Uber drivers, while the gig economy has taken off with a workforce estimated at 10 million in 2015, including about 1.6 million couriers and warehouse workers in e-business alone (Ali Institute, 2016; Fenghuang Wang Finance, 2016). Different levels of the government have set up funding schemes to encourage 'mass entrepreneurship' (Luo, 2016, p. 18): more than 20 provinces now provide loans, rent subsidies, tax reduction to encourage university graduates to set up micro-enterprises, technological incubators, and online businesses (Government of China, 2015). The increase in both public debt and private debt is tantamount to deploying future resources to secure present social peace. Chinese household debt's share of GDP has increased from 30% in 2011 to 60% in 2017 while public debt increased from 15% in 2011 to 38.5% in 2014 (Lee, 2017, p. 33). The politics of credit will become a major arena of struggle as the debt state and the debt society compete for the allocation of credits. The Chinese government's recent national experimentation with using big data to assign social credit rating to all citizens ominously portents the rise of credit as a means of authoritarian control (Chin & Wong, 2016).

3.1.3. Disempowerment

In short, an increasing number of workers face multi-faceted precarity – being excluded from the labour market (laid off, unemployed, or under-employed), dispossessed of their land as a means of social security and subsistence, and forced to incur debts in order to launch their micro-business ventures as self-employed entrepreneurs. As the economic pressures on livelihoods mount, the political space for collective mobilization and self-organizing is also narrowing. The current top leader Xi Jinping, compared to his predecessors Hu Jingtao and Jiang Zemin, has launched exceptionally harsh, widespread and repressive crackdowns on human rights and NGO communities. Reversing Hu's emphasis on social harmony and Jiang's on rule of law and internationalization, Xi has announced zero tolerance for dissent and has demanded total submission, both at the elite and grassroots levels, in the media and education arenas. Anti-corruption campaigns are used selectively to target his political opponents at the top. Arrests and imprisonment of rights lawyers and labour-NGO activists have had chilling effects on worker capacity, just as some labour activists have begun taking bolder action beyond cellular and legal mobilization.

3.2. Precariats' struggles for livelihood

Since around 2010, economic downturn, plant relocation and restructuring have contributed to a rising trend of strikes in the formal sector. In the wake of some high-profile strikes in foreign-invested companies making global consumer products, such as Honda, Foxconn, IBM, and Yue Yuen, some journalists and scholars of Chinese labour saw a tendency of labour empowerment. Their argument, in a nutshell, is that the second generation of migrant workers are more class and rights conscious, social media and technology savvy, demanding union representation in addition to increased compensation, and adopting an extra-legal action repertoire. Yet, closer empirical analysis of the processes and outcomes of these strikes convey quite a different picture. Except on the issue of wage increases, workers did not make any lasting gain in union election, security of employment, and workplace reforms. Also, single factory-based action is still the norm, so is workers' concern to stay within the boundary of the law in their action. There is no evidence to show that second-generation workers are more prone to collective action than the first generation (see Lee, 2016).

A critical and new development was emerging around the time of these attention-grabbing strikes but away from the media limelight. In the past five years, a dozen or so grassroots NGOs, after years of providing individual rights-based legal assistance to workers, sought to augment their impact by mobilizing workers to undertake worker-led collective bargaining with their employers. With the financial support of labour groups outside of China and legal advice of rights lawyers within China, daring NGO activists built networks of worker activists across factories and recruited cross-class participation by students and academics in sustaining strikes. NGO activists coined a new term 'labour movement NGOs' to distinguish themselves from their former selves as service-providing NGOs. They provided moral, legal, and training support to striking workers and, most impressively, liaised worker leaders from different factories to share their experience in bargaining with employers and organizing workers. 42 strikes have been documented in South China between 2011 and 2014, involving 11 labour movement NGOs (Chunyun, 2016; see also Clover, 2016). However, in late 2015, Xi Jinping's government reacted by arresting key labour-NGO leaders and orchestrated smear campaigns against them and their organizations on national television, stifling the confidence and capacity of a budding worker movement. It is uncertain whether repression under the political new normal will end up thwarting or radicalizing these NGOs.

Even as the political space for grassroots NGOs is curtailed, self-mobilized worker struggles have continued. A new tendency is that workers' demands have increasingly turned towards issues of pension, housing, and livelihood, or the social reproduction of labour. As the first generation of migrant workers approach retirement age, they have become more vigilant about employers' making the legally required contributions to their pension and housing funds. Rustbelt state sector workers newly laid off by the state's call to reduce overcapacity also demanded en masse for the state to protect their livelihood and retirement. For informal workers who occupy the blurred boundaries between capital and labour, employed and self-employed, their demands are framed and experienced broadly as a crisis of livelihood. For instance, in 2015, a wave of taxi driver strikes hit major cities in coastal and interior provinces due to the competition of on-demand app-based car services. Even though taxi drivers are self-employed – own their means of production (taxis), pay gasoline, car insurance, and maintenance, and are not employees of taxi companies – they have to pay a fixed 'membership fee' every month to their company in order to participate in this semi-monopolistic industry. The competition of on-demand drivers threatens taxi drivers' 'livelihood' and 'survival', terms they used to describe the reasons for their strikes (Xinhuanet, 2015).

In cities, government encroachment on the use of the 'urban commons' is increasingly depriving precarious workers of a crucial resource for their subsistence economy in the Chinese cities. As in Indian cities (Goldman, 2015, p. 153), government agencies are converting public lands (lakes, bylanes, pavements, marketplaces, bus hubs, parks, etc.) into speculative real estate. Mobilizing the ideology of making 'global cities', these local governments are increasingly displacing the rag pickers, drain cleaners, scrap collectors, cart vendors, porters, and tea and snack sellers to make way for financial speculative capital.

Street vendors' clashes with *chengguan,* a para-police force first set up in the late 1990s, at times turned violent and escalated into mass protests involving local residents resentful of official brutality (Swider, 2015a). In Zengcheng in 2011, a scuffle between a pregnant street vendor and the *chengguan* turned into several days of riots by migrant informal workers who burned government offices and destroyed police cars. On a smaller scale, violent clashes erupted in 2013 between citizens and police after the death of a watermelon vendor who was attacked by the *chengguan* in Linwu, a city in Hunan province (Jacobs, 2013). Such a dynamic is reminiscent of the unanticipated consequences of a street vendor's self-immolation in Tunisia in January 2011 which touched off a powerful political tsunami throughout the Arab world.

Another scenario is also possible. While precarious worker struggles have the potential to be more violent, volatile, and less institutionally incorporated, workers can easily become more atomized and acquiescent. This is so not just because of the frequent change of jobs, depriving them of stable social relations and spatial concentration and communication. As workers cobble together various sources of income and resources, their interests (whether based on market or production) are also differentiated and fragmented (Wang, 2016). Finally, we cannot underestimate the responsiveness of the Chinese state. Despite its autocratic politics, it has a track record of weathering many a socio-economic crisis by responding to social discontent with policy innovations in order to maintain social stability. As it has done in the past in both rural and urban China, the state has been compelled by popular unrest to gradually develop and strengthen social and welfare policies to protect the livelihood of the most vulnerable citizens.[4] As the gig service-oriented economy grows, boundaries of labour and capital blur, and livelihood pressures increase for the general citizenry, the regime may be compelled once again to find policy solutions.

4. Conclusion

Defining precarity as relational struggles, this paper traces a shift in the most salient terrain of contestation from recognition to regulation and now social reproduction of labour, as China evolved from a state socialist political economy to one of a high-growth market economy and then to a new normal of slow growth, overcapacity and enhanced authoritarianism. The driver of precarity in each period also differs, shaping workers' interests, capacities and claims.

Conceptualizing precarity as multidimensional relational struggles rather than a set of fixed attributes may also contribute to understanding precarity comparatively. For instance, are China's workers more precarious than those in India and Brazil? Judging from employment opportunities and wages, there is no denying that China's capitalist boom has lifted millions out of absolute poverty (declining from 84% to 16% of the population between 1981 and 2005). Rural and urban workers have experienced substantial and real improvements in many socio-economic indicators of development, compared to India and Brazil. For instance, despite having the highest poverty rate among the three countries in 1981, China's proportional rate of poverty decline from 1981 to 2015 was 6.5% per annum, compared to 3.2% in Brazil and 1.5% in India (Ravallion, 2011). Yet, poverty reduction does not nullify the existence and politics of precarity, if only because workers everywhere experience and act on precarity in relative and relational ways. India's informal workforce may be larger (82% of non-agricultural workers) and abject poverty more entrenched than in China, but the political space for precarious workers to develop their organizing capacity is much larger. Self-employed workers in India have innovatively sought recognition from the state through the issuance of an 'informal worker identity card', access welfare through trade-based tri-partite 'welfare boards', fight for piece-rate based minimum wage regulations, and even form cooperatives, all resulting in expanded definitions of 'worker' and 'exploitation' (Agarwala, 2016). Precarity, and the struggles emanating from and against it, should, therefore, be theorized in historical, cultural, and context-specific terms.

Notes

1. These concepts are drawn from feminist theories; see especially Fraser (2000, 2016).
2. The supply of student interns as a source of precarious labour has resulted from the commodification of vocational education and the collusion between local government and powerful multinational corporations. Vocational schools have been privatized since the late 1990s and receive equipment, trainers, and funding in return for the internship programmes. Local governments compete with each other to lure big investors like Foxconn to move to their localities, and promise companies a steady supply of interns, see Su (2011).
3. For ethnographic depictions, see Guang (2005), Loyalka (2012), Pai (2012).
4. For the spread of social assistance programme in the global south, including China, in the past two decades, see Harris and Scully (2015).

Acknowledgements

I am grateful to Jan Breman, Alf Nilsen, Mark Selden, Goran Therborn, Karl von Holdt, and two anonymous reviewers for their comments. Song Qi's meticulous research assistance has been indispensable.

Disclosure statement

No potential conflict of interest was reported by the author.

References

Agarwala, R. (2016). Redefining exploitation: Self-employed workers' movements in India's garments and trash collection industries. *International Labor and Working-Class History*, *89*, 107–130.

Ali Institute, Beijing Jiatong University (阿里研究院菜鸟网络). (2016, May). 全国社会化电商物流从业人员研究报告 *[Research report on workers in the ecommerce and logistics section in China]*. Retrieved from http://i.aliresearch.com/img/20160505/20160505154633.pdf

All-China Federation of Trade Unions. (2011). 劳务派遣调研报告,当前我国劳务派遣用工现状调查 *[Report on working conditions of dispatch workers in China]*. Retrieved from http://www.waljob.net/article/6225.html

APCO Worldwide. (2015, November 13). *The 13ᵗʰ five-year plan: Xi Jinping reiterates his vision for China*. Retrieved from http://www.iberchina.org/index.php/evolucicona-contenidos-29/1258-el-13-plan-quinquenal/http://www.iberchina.org/files/13-five-year-plan.pdf

Arrighi, G. (2008). *Adam Smith in Beijing: Lineages of the 21st century*. London: Verso.

Bradsher, K. (2012, March 5). In China, sobering signs of slower growth. *New York Times*. Retrieved from http://www.nytimes.com/2012/03/06/business/global/in-china-sobering-signs-of-a-slower-growth.html

Chan, J. (2017). Intern labor in China. *Rural China*, *14*, 82–100.

Chan, J., Pun, N., & Selden, M. (2013). The politics of global production: Apple, Foxconn and China's new working class. *The Asia-Pacific Journal*, *28*(2), 100–115.

Chan, J., Pun, N., & Selden, M. (2015, September 7). Interns or workers? China's student labor regime. *Asia-Pacific Journal*, *13–36*(2), 1–25. Retrieved from http://apjjf.org/-Jenny-Chan/4372

Chin, J., & Wong, G. (2016, November 28). China's new tool for social control: A credit-rating for everything. *The Wall Street Journal*. Retrieved from http://www.wsj.com/articles/chinas-new-tool-for-social-control-a-credit-rating-for-everything-1480351590

China Labour Support Network. (2014). 探索 中国劳务派遣工处境及改善方向 *[Preliminary study on China's dispatch worker condition and ways for improvement]*. Retrieved from http://www.worldlabour.org/emg/files/Dispatch%20labour%20preliminary%20study%s0(chiese%20report).pdf

Chuang, J. (2015). Urbanization through dispossession: Survival and stratification in China's new townships. *Journal of Peasant Studies*, *42*(2), 275–294.

Chunyun, L. (2016). *Unmaking the authoritarian labor regime: Collective bargaining and labor unrest in contemporary China* (Unpublished doctoral dissertation). Rutgers University, New Brunswick.

Clover, C. (2016, January 11). China police arrest activists in campaign against labour unrest. *Financial Times*. Retrieved from http://www.ft.com/cms/s/0/7ae19510-b85e-11e5-b151-8e15c9a029fb.html#axzz42iTl2dqh

Dan, M. (2015). 北京市住家家政工的劳动过程分析 [A labour process analysis of domestic labour in Beijing]. *Chinese Workers*, *2*, 18–22.

Duhigg, C., & Bradsher, K. (2012, January 21). How the US lost out on iPhone work. *New York Times*. Retrieved from http://www.nytimes.com/2012/01/22/business/apple-america-and-a-squeezed-middle-class.html?hp

Fan, L. L., & Xue, H. (2017). The self-organization and the power of female informal workers: A case study of the cooperative production team in the garment industry in the Yangtze river delta. *Rural China*, *14*(1), 61–81.

Fenghuang Wang Finance. (2016, August 15). 去年中国共享经济市场规模约1.95万亿 从业人员近千万 *[The sharing economy reached 1.95 trillion yuan and 10 million employees last year]*. Retrieved from http://finance.ifeng.com/a/20160815/14747459_0.shtml

Fraser, N. (2000, May-June). Rethinking recognition. *New Left Review*, *3*, 107–120.

Fraser, N. (2016, July–August). Contradictions of capital and care. *New Left Review*, *100*, 99–117.

Friedman, E. (2014). *Insurgency trap: Labor politics in postsocialist China*. Ithaca: Cornell University Press.

Gallagher, M. E. (2014). China's workers movement and the end of the rapid-growth era. *Daedalus*, *143*(2), 81–95.

Goldman, M. (2015). With the declining significance of labor, who is producing our global cities? *International Labor and Working-Class History, 87*, 137–164.

Government of China. (2015, June 11). 国务院关于大力推进大众创业万众创新 若干政策措施的意见 [*Policy suggestions by the state council on promoting mass entrepreneurship*]. Retrieved from http://www.gov.cn/zhengce/content/2015-06/16/content_9855.htm

Grada, C. O. (2011). Great leap into famine: A review essay. *Population and Development Review, 37*(1), 191–202.

Guang, L. (2005). Guerrilla workfare: Migrant renovators, state power, and informal work in urban China. *Politics and Society, 33*(3), 481–506.

Harris, K., & Scully, B. (2015). A hidden counter-movement? Precarity, politics, and social protection before and beyond the neoliberal era. *Theory and Society, 44*(5), 415–444.

Heilmann, S. (1993). The social context of mobilization in China: Factions, work units and activists during the 1976 April Fifth Movement. *China Information, 8*(3), 1–19.

Huang, G. Z. (2015). 城市摊贩的社会经济根源与空间政治 [*The socio-economic roots and spatial politics of city street vendors*]. Beijing: The Commercial Press.

Huang, Y. (2012). 工厂外的赶工游戏--- 以珠三角地区的赶货生产为例》社会学研究 [Rush order game beyond the factory: The case of rush production in the Pearl River Delta]. *Sociological Research, 4*, 187–203.

Hung, H. (2015). *The China boom: Why China will not rule the world.* New York, NY: Columbia University Press.

IIEC (Institute of International Economic Cooperation, Ministry of Commerce) (商务部国际贸易经济合作研究院). (2015, June). 中国家政服务行业发展报告 [*Report on the development of China's domestic service sector*]. Retrieved from http://images.mofcom.gov.cn/fms/201509/2015092916430692.pdf

Jacobs, A. (2013, July 20). Death of watermelon vendor sets off outcry in China. *New York Times.* Retrieved from http://www.nytimes.com/2013/07/21/world/asia/death-in-china-stirs-anger-over-urban-rule-enforcers.html?_r=0

LANDESA. (2012, April 26). *Summary of 2011 17-Province survey's findings: Insecure land rights; The single greatest challenge facing China's sustainable development and continued stability.* Retrieved from http://www.landesa.org/china-survey-6/

Lee, C. K. (2007). *Against the law: Labor protests in China's rustbelt and sunbelt.* Berkeley: University of California Press.

Lee, C. K. (2016). Precarization or empowerment? Reflections on recent labor unrest in China. *Journal of Asian Studies, 75*(2), 317–333.

Lee, C. K. (2017). After the miracle: Labor politics under China's new normal. *Catalyst, 1*(3), 93–115.

Lee, C. K., & Zhang, Y. H. (2013). The power of instability: Unraveling the microfoundations of bargained authoritarianism in China. *American Journal of Sociology, 118*(6), 1475–1508.

Li, K. (2013, May 26). Li Keqiang expounds on urbanization. *China.org.cn.* Retrieved from http://www.china.org.cn/china/2013-05/26/content_28934485.htm

Li, K. (2016). Rising inequality and its discontents in China. *New Labor Forum, 25*(3), 66–74.

Lim, B. K., Miller, M., & Stanway, D. (2016, March 1). Exclusive: China to lay off five to six million workers, earmarks at least $23 billion. *Reuters.* Retrieved from http://www.reuters.com/article/us-china-economy-layoffs-exclusive-idUSKCN0W33DS

Lin, J. Y. (n.d.). *Demystifying the Chinese economy* (Unpublished manuscript). Retrieved from http://siteresources.worldbank.org/DEC/Resources/84797-1104785060319/598886-1104852366603/599473-1223731755312/Speech-on-Demystifying-the-Chinese-Economy.pdf

Loyalka, M. D. (2012). *Eating bitterness: Stories from the front lines of China's great urban migration.* Berkeley: University of California Press.

Luo, W. S. (2016, March 3). Uber to get more women on the road. *China Daily.* Retrieved from http://www.chinadaily.com.cn/kindle/2016-03/03/content_23723294.htm

Naughton, B. (2007). *The Chinese economy: Transitions and growth.* Cambridge, MA: MIT Press.

Naughton, B. (2009, May). Understanding the Chinese stimulus package. *China Leadership Monitor, 28*), Retrieved from http://media.hoover.org/documents/CLM28BN.pdf

Pai, H. (2012). *Scattered sand: The story of China's rural migrants.* London: Verso.

Pang, I. (2016). *Precarity as a legal construction: lessons from the construction sector in Beijing and Delhi.* (Unpublished manuscript). Brown University, Providence.

Perry, E. J. (1994). Shanghai's strike wave of 1957. *The China Quarterly, 137,* 1–27.

Perry, E. J. (1996). Labor's love lost: Worker militancy in communist China. *International Labor and Working-Class History, 50,* 64–76.

Ravallion, M. (2011). A comparative perspective on poverty reduction in Brazil, China and India. *The World Bank Research Observer, 26*(1), 71–104. Retrieved from https://openknowledge.worldbank.org/handle/10986/13499

Rungain Think Tank of Entrepreneurship. (2016). 荣硅创业智库 *(2016)* 中国创新创业 *2015* 年度报告 *[Report on China's innovative entrepreneurship]*. Retrieved from http://www.zyfsgs.com/gongzuobaogao/2017/0613/246266.html

Selden, M. (1993). *The political economy of Chinese development.* Armonk, NY: M. E. Sharpe.

Solinger, D. J. (2009). *State's gains, labor's losses: China, France, and Mexico choose global liaisons, 1980–2000.* Ithaca: Cornell University Press.

Standing, G. (2011). *The precariat: The new dangerous class.* London: Bloomsbury.

Su, Y. (2011). Student workers in the Foxconn empire: The commodification of education and labor in China. *Journal of Workplace Rights, 15*(3–4), 341–362.

Swider, S. (2015a). Reshaping China's urban citizenship: Street vendors, *Chengguan* and struggles over the right to the city. *Critical Sociology, 41*(4–5), 701–716.

Swider, S. (2015b). *Building China: Informal work and the new precariat.* Ithaca: Cornell University Press.

Thompson, E. P. (1963). *The making of the English working class.* New York, NY: Vintage.

USCBC (US-China Business Council). (2010, July 1). China's priorities for the next five years. *China Business Review.* Retrieved from http://www.chinabusinessreview.com/chinas-priorities-for-the-next-five-years/

Walder, A. G. (1986). *Communist neo-traditionalism: Work and authority in Chinese industry.* Berkeley: University of California Press.

Wan, X. (2015, September 14). 农民工非正式就业者的生存状态与网络支持. *[Informal migrant workers' livelihood conditions and support networks]* (Unpublished manuscript). Retrieved from http://www.taodocs.com/p-32788539.html

Wang, J. H. (2016, October). *Dependent government-corporate relations, local employment and labor contentions in inland city* (Unpublished manuscript).

White, L. T. (1976). Workers' politics in Shanghai. *Journal of Asian Studies, 36*(1), 99–116.

Whyte, M. K. (ed.). (2010). *One country, two societies: Rural-urban inequality in contemporary China.* Cambridge, MA: Harvard University Press.

Whyte, M. K. (2014). Soaring income gaps: China in comparative perspective. *Daedalus, 143*(2), 39–52.

Wong, C. H. (2016, November 29). China looks to loosen job security law in the face of slowing economic growth. *Wall Street Journal.* Retrieved from http://www.wsj.com/articles/china-looks-to-loosen-job-security-law-in-face-of-slowing-economic-growth-1480415405

Wu, J. (2017). Migrant citizenship regimes in globalized China: A historical-institutional comparison. *Rural China, 14*(1), 128–154.

Xinhuanet. (2015, January 7). 多地现出租车停运事件,仅是"专车""黑车"惹的祸? *[Taxi boycotts in many cities: Who created the problem?]*. Retrieved from http://news.xinhuanet.com/mrdx/2015-01/07/c_133901604.htm

Zhang, L. (2015). *Inside China's automobile factories: The politics of labor and worker resistance.* New York, NY: Cambridge University Press.

Social mobilizations and the question of social justice in contemporary Russia

Karine Clément

ABSTRACT
Notwithstanding stereotypes of Russian apathy, long-term field research reveals that there have always been grassroots and labour protests in post-Soviet Russia, even as the shock of ultraliberal reforms led to mass precariousness and social disorientation. However, the social mobilizations that do occur are scattered, weakly publicized and mostly small-scale. This paper conceptualizes them as 'everyday activism', that is, an activism embedded in everyday life experience and pragmatic sense. Only recently, and in a paradoxical relation to the populist and patriotic Kremlin discourse, some new trends have emerged towards other popular variants of the new discourse that includes social equality claims and what the paper calls 'social critical' populism. However, this populism from below does not automatically lead to mass mobilization, although it provides the necessary background for it.

The dramatic social shock and harsh neoliberal reforms that destabilized and traumatized millions of Russians after the fall of the communist system have not led to mass mobilization. It is striking that in a country where social and labour precariousness is so widespread, claims for social justice are not on the agenda either of the liberal political opposition or of grassroots social movements. Social disorientation has been pervasive for so long that most people had no sense of belonging to any social groups and could not even imagine the larger society through any social coordinates. Only recently, together with the economic crisis that began in 2014 and the awakening of national consciousness, has social justice seemed to emerge as issue for ordinary people, in their everyday conversations, and sometimes in their explicit social claims.

I interrogate the social justice question because it is the big void in Russian political and intellectual debates. Social justice is here taken in the sense of socio-economic justice, including claims to welfare entitlements and a sense of exploitation and domination (not in the ethical or moral sense of fairness). I raise this question because of the delegitimization of welfare claims in post-Soviet society and because of Russia's exceptional levels of socio-economic inequality (Novokmet, Piketty, & Zucman, 2017). The second reason is that, after two decades of trivialization of socio-economic injustice (that is taken as 'normal' in the transition from communism to capitalism), some people are beginning to speak about it. Recent studies indicate the development of a social consciousness, which had long disappeared after the breakdown of the Soviet system. However, and this is the third reason for addressing the question, socio-economic inequalities remain almost invisible in the public sphere and are hardly manifested in protest movements. It is all the more surprising that in many other parts of the world, particularly after the global financial crisis of 2008, protest movements voiced the question, whether they called it 'anti-austerity' (Angelovici, Dufour, & Nez, 2016; Della Porta, 2015), 'social justice' (Glasius & Pleyers, 2013), or 'anti-oligarchic' (Gerbaudo, 2017).

This paper tracks the emergence of the social justice question in post-Soviet mobilizations and, partly, in everyday settings involving less mobilized people. The analysis relies mainly on primary data from field research on labour relations (Clément, 1999, 2004, 2008), grassroots social initiatives (Clément, 2013a; Clément, Demidov, & Miryasova, 2010), and everyday nationalism.[1]

The paper first outlines the manner in which the sociol-economic system developed after 1989, a critical component in the radical transformation of social power relations that gave rise to socio-economic inequality while preventing protest. To grasp the conditions for a social justice movement, I analyse in greater detail three instances of local mobilizations where issues of socio-economic justice were raised. In the second part of the paper, I examine the relationship between so-called 'political' protests (for example, the movement 'For Fair Elections' in 2011–2012) and socio-economic ones. The last section demonstrates the spread of a sense of socio-economic injustice among a large part of the population. It attempts to explain which changes in the social power relations might have led to a growing sense of social inequality.

Social movements facing two waves of neoliberal reforms

Radical neoliberal capitalist reforms undertaken just after the breakdown of the communist regime led to social disorientation, impoverishment, and precarization of a large majority of the population. The neoliberal course has not changed up to the present but what has changed is the rhetorical packaging of the reforms. The ultraliberal tone of Boris Yeltsin's 1990s has been replaced by the populist and patriotic discourse of the Vladimir Putin era.

The trauma of the 1990s

In the 1990s, Russia underwent an economic 'shock therapy' in which ultraliberal reforms were ruthlessly implemented in terms of a structural adjustment model required by the International Monetary Fund. This led to unprecedented economic decline and social breakdown. As a result, real incomes fell by 43.7% from 1991 to 1992, according to official Russian state statistics. Wages fell even more drastically, while wage arrears grew explosively over the 1990s.

The result was the brutal impoverishment of most people who depended on wages and social payments. Judah (2013) described the period as 'the wild nineties', 'a synonym in Russian for a decade that left practically every family with stories of deprivation, unpaid wages, economic humiliation and diminished status' (p. 13). The welfare system crashed after the collapse of the Soviet Union, and most people found themselves deprived of social security and uncertain of what tomorrow would bring. The market ideology urged people to make it on their own and not to rely on state care.

Thus, the first stage in the formation of Russian capitalism led to a sharp economic downturn and the impoverishment of the majority of the population, while a minority of private property owners and power holders succeeded in privatizing or controlling economic assets and monopolizing national wealth. However, instead of revolting and rising up against power holders and oligarchs, people rejected politics and activism and retreated into their private lives and households.

Why did ordinary people remain so passive in the face of the degradation of their conditions? My conclusions draw on ethnographic insights from the labour world in the 1990s and are congruent with other studies (Ashwin, 1998; Burawoy, 1999). The workers' weak resistance is understandable when one takes into account their dependence on individual survival strategies caused by their atomization, precarization, and social destabilization. As a whole, I condense these demobilizing dynamics into the concept of desubjectivation, which is the tendency towards the loss of some sense of the self and agency (Clément, 1999). This concept embraces the process of precarization that undermines the possibility of mass mobilizations. Although precarization is extensive in neoliberal capitalism globally, widely analysed in the literature (Boltanski & Chiapello, 1999; Breman & van der Linden, 2014; Sennett, 1998; Standing, 2011), it was possibly nowhere so widespread and all-pervasive as it was in Russia in the 1990s. Social, labour, and material precariousness destabilized people, broke solidarities, and led to survival strategies, including holding multiple jobs, engaging in subsistence and petty commodity production, and experiencing disqualification, despair, and exhaustion. Most workers struggled to live or to compete with each other to achieve better positions. 'Each one for oneself', as workers constantly repeated in interviews at the time.

The concept of desubjectivation also embraces the dominant neoliberal or consumerist ideology that led to self-depreciation. Impoverished and precarized people tended to blame themselves instead of the system, painfully enduring privations. Blue-collar workers or industrial bottom workers were among the first victims of precarization caused by deindustrialization, the loss of their previous symbolic significance, and the weak mobilizing potential of their organizations. The relative quiescence of blue-collar workers results from their – successful or unsuccessful – adaptation to the social transformations produced by neoliberal reforms and the marketization and commodification of life. Given the degree and scale of these transformations, adaptation often amounted to complete human flexibility, the ability to bend down and distort oneself without breaking. Some were broken by this (e.g. men's life expectancy fell dramatically during the 1990s), while others lost themselves in the never-ending course of adapting to changing conditions. Most of them lost points of reference and orientation, experiencing troubles identifying themselves and identifying the society they lived in. In interviews conducted during the 1990s, many workers used derogatory terms to describe themselves: 'a small screw in the soulless machine', 'a nothing', 'unneeded people', 'cattle', or 'slaves'.

Nevertheless, social struggles did sometimes occur in the 1990s, especially in the form of labour protests. These mostly broke out in a spontaneous and disorderly fashion, generally over the issue of unpaid wages (Robertson, 2010), and were not part of any mass movement. The first indication of a revival of the workers' movement occurred in the late 1990s when some workers had recovered from the initial economic downturn and were no longer willing to endure wage delays or non-payment. A wave of labour struggles in 1998–1999 was also a reaction to the dismantling of production structures. The new owners of privatized firms had often bought state enterprises cheaply in order to sell everything they could in the short term, without any consideration for production development. Acting against the dilapidation of the equipment and machines, workers sometimes organized themselves to control plants. There were several cases of workers occupying plants, one of the longest being the occupation of the Vyborg Cellulose Paper Combine (1998–1999). Another large protest movement, which roused solidarity, involved miners' strikes and blockades of the Trans-Siberian railway in May–June 1998, in which not only miners participated but also railway machinists, teachers, and municipal service workers in several towns. Workers first protested against wage arrears and then demanded Yeltsin's dismissal. Their action took a political tone and expressed discontentment with 'wild capitalism', 'cleptocracy', criminals, and nasty oligarchs. While the common adversary (Yeltsin) and a shared centre of attention (the miners' protest camp at the Humpback Bridge in Moscow) could have triggered a mass protest movement, by August 1998 the ruling elite undermined its possibilities by devaluating the rouble and preparing a new leadership. These steps reactivated industrial growth, while Putin took on the popular demands for 'struggling against the oligarchs', 'fighting poverty', and restoring order, the state, and 'the rule of the law'.

The economic revival of the 2000s and the development of grassroots social movements

The improvement of the material and economic situation as well as an increasing satisfaction with the Kremlin's new priorities in the 2000s led the labour struggles of the 1990s to fade away. In the new millennium, real wages and pensions recovered, unemployment declined, and poverty fell by half. Many people experienced relief from the day-to-day struggle for survival and recovered a relatively reliable economic basis. However, at this point, Putin's government began a new phase of neoliberal reforms, this time aimed at restructuring the social welfare system.

In 2001, a flat income tax rate of 13% was adopted. A new labour code implemented in 2002 improved employers' positions to the detriment of employees, especially in terms of the rights for establishing independent trade unions and organizing strikes. In the later 2000s, a reform of the legislation on housing and urban and ecological issues increased the cost of residents' utilities and housing maintenance while opening the path to the privatization of communal services, housing, and land. Under the influence of the World Bank and the World Trade Organization, the government chose price deregulation and privatization as prime criteria for its decisions.

As people had recovered from the shock of the 1990s, these new anti-social reforms launched by Putin's government gave rise to protest, beginning in the mid-2000s. When Putin's government attacked the social benefits system in late 2004, it was confronted with one of the most massive protest movements post-Soviet Russia has known (Clément, 2013b). The social movement during the winter of 2004/2005 opposed the monetization of in-kind social benefits (*lgoty*) that actually cut the social benefits of a number of specific professional categories and in practice affected most of the population, particularly retirees and veterans, school children, students, and the disabled, Chernobyl liquidators, Great North workers, and victims of political repressions. Protest actions began on a very small scale, following altercations on buses and trolleys when retirees objected a new rule by which they had to pay for their tickets. The news spread like wildfire, fanned by feelings of indignation, injustice, and contempt.

The movement quickly gained traction, mobilizing more than 500,000 people in 97 towns and 78 regions across the country.[2] The national campaign ultimately achieved a partial repeal of the reform, a rare occurrence in contemporary Russia.

With this concession, the social policy course seemed to have been corrected in order to fit people's aspirations and concerns. Federal programmes in health, housing, education, and natality were launched with loud publicity. For a short time, these programmes had positive results, improving access to affordable housing and increasing the wages of education and healthcare workers. When Russia was hit by the global economic crisis of 2008–2009 and a recession in early 2014 (because of a fall in oil and gas prices, as well as sanctions and counter-sanctions declared after the Russian annexation of the Crimea), it halted the improvement of households' social situation and living conditions. Currently, the problems of wage arrears, the diminution of real wages, and impoverishment have made a dramatic comeback, while social inequality has been reinforced.

Despite the worsening of the socio-economic situation and the rise of social inequality, no massive movement has emerged to raise questions of social justice. The most widespread form of protest is diffuse and a form of local grassroots mobilization that is rooted in the daily lives of its participants and seeks to address particular and narrow social problems (the closure of schools or hospitals, the increase in transport or communal charges, problems of urban construction, delays in the payment of wages). The lack of a unifying cause could be one of the difficulties that have hindered the development of a nationwide movement. Another problem is the weak publicity that grassroots movements get from Moscow-centred media and intellectuals, many of whom criticize grassroots initiatives for their narrow-mindedness. A third obstacle is the lack of mobilizing structures due to the high degree of atomization in society and a loss of the sense of social belonging.

One of the few nationwide organizational structures is the trade unions, but they are not very powerful in mobilizing. The Russian trade union movement remains dominated by former Soviet official unions, renamed the Federation of Independent Trade Unions of Russia (FNPR), which generally collaborate with management, thus contributing to the delegitimization of trade unions. The alternative or free trade unions frequently focus on the defence of labour rights and a confrontation with employers, but they face difficulties in gaining recognition and support among workers.

A dynamic towards more consolidation between small-scale and local grassroots or labour movements as well as towards social justice claims began developing in 2007–2008. There was a quantitative rise in local grassroots and single-issue movements in a number of towns and cities. In some cases, this led to a dynamic of interconnections, and sharing that produced larger multi-issue, city-wide movements with common claims. One such case was the Kaliningrad protest movement of 2009–2010 that achieved the dismissal of the regional governor (Clément, 2015a) and brought together such different interests as a movement of car drivers and importers, a mobilization against a hospital's closure, an organization of small shopkeepers, a retirees' campaign, and a rally of a local airline's employees.

At the end of the 2000s, besides the development of urban movements, another process of consolidation took place in the realm of labour relations. In contrast to the relative quiescence of the early 2000s, 2007–2008 were marked by a strike wave that extended into many industries and regions and raised labour conditions and pay as issues. This labour mobilization was made possible by a renewal of the trade union movement and by the stable economic growth since 2000. Indeed, beginning with the strike at Ford near St. Petersburg in November to December 2007, strikes burst out in many other enterprises with the help of newly created alternative unions. At the same time, these alternative trade unions achieved a long-aspired for unification and joined the Confederation of Labour of Russia (KTR).

The 2008 global economic recession changed the character of this mobilization, pushing defensive claims against delayed payments and layoffs to the forefront, but it neither stopped the mobilization process nor the progress of unionizing new sectors. Street protests were quite radical in 2009/2010, addressing claims at the level of the state, since employers were not responsive to the crisis. One of these protests was the blockade of a federal highway in the industrial monotown of Pikalyovo (Leningrad region). Workers in the metallurgical factories, together with their families and with the support of the local branch of the traditional union, the FNPR, took to the street to call for Putin to intervene in their conflict with the factories' owners. The blockade and its media resonance led to Putin's public and mediatized intervention in the conflict.

Three movements voicing social inequality

Social movements seldom addressed issues of socio-economic inequality and justice. There were some exceptions, however, mostly in movements that emerged from grassroots initiatives in areas well removed from wealthy centres and in mobilizations that embraced labour issues. Below, I track some of these grassroots mobilizations, which vociferously raised social inequality as an issue, in order to understand their specificities. Among the many cases of grassroots social mobilization I have studied in Russian regions since 2005, I choose to concentrate on three cases: the labour conflict at the Ford plant in Vsevolozhsk, Leningrad region, at the end of 2007 (field research conducted in February 2008); the workers' and inhabitants' movement of 2010/2011 to save the industrial mono-town of Rubtsovsk, Altai, Krai (field research in July 2011); and the urban movement of 2012 in Astrakhan that opposed the Astrakhan city authorities (field research in April 2012).

In all three cases, core participants had relatively low social positions and shared a rather homogeneous socio-economic status. At Ford, the participants in the November–December 2007 strike were all subaltern workers with wages slightly above the average in the national car industry. They went on strike to enhance their labour conditions and raise their salaries. In Rubtsovsk, most of the protesters were low-paid, unpaid, or former workers of the Altai Tractor Plant. After two relatively successful collective hunger strikes in May and June 2010 to demand overdue salaries, many ex-workers joined the dwellers' committees in a common struggle against an increase in communal charges, expanding into massive rallies in 2011. In Astrakhan, a network of various grassroots initiatives had developed, including a network of neighbourhood committees, the movement by drivers/owners of collective taxis, the movement of street market workers, and the local Defence labour union. In March–April 2012, they all participated to support the oppositional candidate against the 'corrupt' local government that was contemptuous of the needs of ordinary Astrakhanys.

What differentiated the protesters in Vsevolozhsk, Rubtsovsk, and Astrakhan from other protesters was both their remoteness from a prosperous centre and their local entrenchment. Vsevolozhsk (64,000 inhabitants) is a small, emerging industrial city located on the St. Petersburg periphery, while Rubtsovsk (145,000 inhabitants) is a middle-size industrial city in crisis, located far from the regional centre (Barnaul). Astrakhan (530,000 inhabitants) is a regional capital in economic crisis, located on the southern borders of the country. The remote location of all three sites influenced the self-definition of their inhabitants: people tended to depict themselves as deprived of attention from the central authorities and from people from large cities.

Partly because of their social position, grassroots activists from the three cases were more likely than in other places I have studied to express a sense of collective identity and identify themselves with concrete social groups: 'the working people', 'the common people', or 'the working class'. At Ford, they depicted themselves as 'blue-collar workers' or even as 'working class' who stood in opposition to 'office workers', 'management', and the 'administration'. The Rubtsovsk workers and apartment building inhabitants mainly depicted themselves as 'the lower classes of the people', people deprived of power, social security, and respect. 'Relations need changing. Relations towards people, working people, working class … These new Russians don't even consider us as human beings', said a worker at a water treatment plant who tried to oppose layoffs. Astrakhan activists shared the same feeling of belonging to the 'little people', 'ordinary people', or 'poor people' who were 'from the bottom', as opposed to the 'thieves', 'bandits', 'well-offs', and 'capitalists' who were 'above'.

It is worth noting that grassroots movements from these cases stood out for bringing forward the labour issue, while many other grassroots movements tended not to address this issue, at least in their public agenda. Indeed, labour is not considered a political issue in neoliberal capitalist Russia, but an individual and even personal activity. Unlike many other grassroots activists who focus their blame exclusively on the government or local officials, activists from Astrakhan, Rubtsovsk, and Vsevolozhsk blamed 'capitalists' and 'oligarchs' for their problems and saw themselves as victims of 'capitalistic logic'.

However, these local and grassroots social justice struggles also had their limits: they did not articulate over-arching social justice project that could appeal to movements more widely. Activists fought against unpaid wages, the increase in price of services, the closure of factories, or the local government: they did not fight for social justice as such. This was largely due to their sense of power-lessness in tackling general problems as well as to their reluctance of dealing with politics. Most of the local grassroots activists consciously avoided politics, and some even denied any political engage-ment. The only politics they might recognize was the 'bottom up politics', 'people's politics', 'politics we are doing here in our houses', or 'politics of the small deeds'. This conception of a very localized form of politics limited the potential for local grassroots movements to spread and for the emergence of a national social justice movement.

The disconnection between high-visibility Putin-centred politics and low-visibility social and economic everyday activism

Towards the late 2000s, there were plenty of grassroots initiatives and labour protests developing all over Russia, sometimes unifying into campaigns for one or another common goal. However, the emergent process of consolidation and expansion slowed down when, at the forefront, emerged the movement 'For Fair Elections' (2011–2012). Unlike grassroots initiatives and labour protests, this movement was relatively unified, mostly involving highly educated people in liberal professions from Moscow and St. Petersburg, but also attracting a number of people from more outlying regions or with lower social status (Bikbov, 2012; Erpyleva & Magun, 2014; Gabowitsch, 2016b; Smyth, Sobolev, & Soboleva, 2013; Volkov, 2012). The movement emerged as a protest against electoral fraud in the 2011 legislative elections and quickly acquired an oppositional character, directed in par-ticular against the prospect of a third presidential term for Vladimir Putin. Unlike grassroots initiat-ives and labour protests, the oppositional movement gained high visibility in the media and credibility in intellectual circles. This movement, framed as a political protest against Putin, eclipsed social and economic protests.

I ground my argument for the precluding effect of the 'For Fair Elections' movement on the for-mation of a large-scale social justice movement in four considerations. First, the grassroots initiatives that did join the movement (such as the movement to defend the Khimki Forest near Moscow) faded in the 'anti-Putin' movement and disappeared for a while as grassroots initiatives. Second, the move-ment did not place social, ecological, or economic issues at the top of its agenda, instead raising only narrow political issues (support for or opposition to Putin) and moral considerations (honesty, goodness). Third, it temporarily occupied all the (limited) public space devoted to dissent, narrowing the political imagination of new activists to 'anti-Putin' protest. Four, both pro- and anti-Putin intel-lectuals and media labelled the movement 'the Opposition', a label in which not all the participants recognized themselves and which is not attractive to most of the population.

Towards the end of 2012, the 'oppositional movement' began to shrink, as many people began to resent the uselessness of 'protest for protest's sake', were disappointed by its lack of results, and had difficulties figuring out the meaning of the mobilization. Many quit activism completely. However, the 'For Fair Elections' movement also led to the development of new grassroots initiatives as some participants went into local activism, to 'tackle concrete local problems' and perform 'real deeds' (Zhuravlev, Savelyeva, & Erpyleva, 2014).

Yet, other participants joined the pro-Putin movement, especially after the annexation of the Crimea in 2014 and Putin's expressed policy of resisting 'Western dictates'. There were several mass mobilizations in 2014–2016 explicitly in support of Putin and the annexation. Many commentators quickly labelled them state-controlled mobilizations. However, sociological interviews with participants show a more complex picture and demonstrate the presence of social and economic protest. A field-study (Mandrik, 2017) conducted among participants in pro-Kremlin mobilizations in Moscow and St. Petersburg in 2015 shows that many respondents, while supporting personally Putin and being patriotic-minded, express discontentment with the social and economic situation in the country. They denounce not only the liberal course of the 1990s, but also the current socio-economic course, the weight of liberals in the actual government, and the oligarchs' power. Here is a typical quotation:

> I am dissatisfied with the system, there is such a narrow circle of elite people, while the majority of people lives poorly. [Interviewer: so you are discontent with Putin?] No, why do you say that? I have nothing against him as a person. He is now close to the simple people [...] What I think is that we don't need any capitalism here.

Even among pro-Putin nationalistic organizations, some activists expressed doubts about Putin and 'whether he really struggles for our well-being, for the well-being of the Russian people, and not for the gas and oil nomenklatura' (quotation from an interview with a member of the National Liberation Front, Perm, May 2016).

Thus, we can assume that the 2014–2016 pro-Putin moblizations carried at least partly a protest character. Another feature of these mobilizations is that many participants share a common resentment towards the liberal opposition, which is perceived as contemptuous of the people. Indeed, liberals often use derogatory terms to refer to the patriotic-minded and pro-Putin unenlightened majority, such as *sovki* [people still 'stuck' in the Soviet past] or *vatniki* [literally, 'quilted jackets']. It is not rare in interviews to hear people emotionally asserting that they are not *vatniki*, but worthy citizens. Anyway, by 2017, pro-Putin mobilizations faded, certainly partly because of a worsening economic crisis in the country which confronted large parts of the population with decreasing real incomes, high inflation, and layoffs.

The crisis in the Russian economy has generated a new wave of labour protests, including in poorly unionized or activist sectors such as services, food industries, and the public sector. They primarily address the delayed payment of wages and 'arbitrary management' that manifests in unjustified layoffs or reorganizations. In fact, the number of labour disputes has been constantly growing since 2007 (Biziukov, 2017), mostly taking the form of small-scale street actions in times of austerity. However, these protests remain scattered and poorly organized, and they have not led to any movement for the improvement of the workers' conditions as a whole. Up to now, the labour issue as well as the trade unions continues to be marginal questions in Russian public debates.

Yet, the vitality of grassroots activism is currently high in Russia. It takes many forms and has many different meanings that are not reducible to either/or patterns. In particular, it is not possible to restrict it to pro- versus anti-Putin positions or to dissent versus loyalty. Activism in the post-Soviet area (and surely beyond) is more typically both practically oriented and oppositional, supportive while also resisting (Jacobsson, 2015). It is a form of 'everyday activism', namely one that is embedded in everyday life experience and pragmatism. The concept refers to Scott's (1985) 'everyday resistance', but it takes into account a wider scope of variable forms of activism, not limited to 'hidden transcripts'. Sometimes everyday activism can enter the public sphere, in particular when the discourse of the ruling elite provides a background for it, as is the case with the patriotic and populist discourse of the post-Crimea era. Sometimes it takes the form of a relatively consolidated urban or national movement with overt dissent, as was the case in the 2005 movement against monetization. But at other times it is half-hidden, entrenched in small areas of commonality (such as conversations in stairwells, courtyards, kiosks, and shops). In any case, it remains 'everyday' because people mobilize on the basis of their everyday experience, learning activism while doing it.

As a whole, these different kinds of everyday activism develop largely out of the spotlight: they remain 'hidden', though not by the actors' choice. Although there is a growing literature that attests to the dynamism of activism in Russia, the dominant discourse in the media and the intellectual sphere still conveys an image of a Russian mythic quiescence. This is largely due to the normative position applied by intellectuals (who apply the 'activism' label only to what they consider as 'right' or 'good' activism) as well as to a top-down conception of politics that is focused on Putin and the central government. In order to grasp what is really going on, it seems necessary to look at social dynamics at the bottom with ethnographic sensitivity, stepping away from predefined normative schemas. This approach underlies much recent research on social mobilization or other forms of civic activism in Russia, paying careful attention to the everyday, ordinary, and familiar (Aidukaite & Fröhlich, 2015; Biziukov, 2017; Clément, 2015a, 2015b; Erpyleva & Magun, 2014; Evans, 2015; Gladarev & Lonkila, 2013; Greene, 2014; Koveneva, 2011; Tykanova, 2012; Zhuravlev et al., 2014). Another growing trend questions state-orchestrated or -manipulated movements regarding the multiple meanings that laypeople may place in them (Daucé, Laruelle, Le Huérou, & Rousselet, 2015; Gabowitsch, 2016a; Hemment, 2009; Krivonos, 2015).

Yet, if activism is developing, its scope remains limited. Participants in grassroots and labour protests seldom go beyond their local or specific problems to build extra-local movements. If they have a sense of social injustice, they do not articulate it in a social justice project, in terms of resource redistribution or welfare entitlements. An additional constraint is the difficulty of social orientation and identification that most Russians faced after the breakdown of the Soviet Union. The next part will be dedicated to the study of this problem and the perspective of a new social imaginary.

From desubjectivation to the revival of social consciousness

In the 1990s, surveys as well as ethnographic research indicate a post-Soviet society characterized by a 'difficulty in mastering, managing the social world' (Oushakine, 2000, p. 998; see also Danilova & Yadov, 1997). As shown in the interviews that I engaged in throughout the 1990s and 2000s, people were generally unable to create an image of society for themselves and to place themselves within it. This has recently started to change due, first, to the relatively long experience of economic and labour stability from 2000 to 2008 that gave stability to those who had previously lost self-confidence. Second, the development of a social sense is also enabled by the re-foundation of the mundane and by inhabiting the immediate material environment, a condition for engaging with and making sense of the world and one's life experience. The concept of inhabitation or habitability is what Morris (2016), studying working class people in a Russian monotown, calls 'the striving for mundane comfort and ordinariness' (p. 8). Inhabiting is 'making habitable the inhospitable and insecure space of lived experience' (p. 236). Thanks to the anchorage in one's daily life, some people recover a sense of the social space, which gives them a chance to blame not themselves or individuals for the inability to 'make it', but to raise the question of social inequality instead.

Changes in the official public discourse, the third point identified here, have also been conducive to social reorientation. During the ultraliberal and anti-communist discourses that dominated the 1990s (sustained by media that was closely linked to political authorities and oligarchs), those who needed social protection and labour security were portrayed as old, reactionary, or incompetent people who had failed to adapt to the market and deserved their miserable existence. The tone of the 'democratic' media was particularly disrespectful and ironic while reporting protest actions for the payment of wages: protesters were depicted as 'fools', 'lazy', 'reactionary', 'irresponsible', or 'extremists'. Government officials and liberal intellectuals actively participated in the stigmatization of those who needed or demanded social security from the state. In this manner, the majority of the common folk or post-Soviet ordinary people became 'losers', supposedly by their own fault as they supposedly lacked the personal qualities needed in a new, modern, and democratic capitalist-market era (Danilova, 2014).

Thus, a capitalist-market ideology predominated in the 1990s. Social inequality was treated as good, and equality was associated with the Soviet *uravnilovka* (equalization or levelling). Because of the rejection of state-imposed communist ideology, criticism of capitalism was tabooed and class language was rejected. Claims on the state for welfare were regarded as a mark of individual failure and paternalism.

Yet the tendency seems to be reversing, with the new populist and patriotic discourse developed by the Kremlin in reaction to the massive protests in 2005 against the social welfare 'anti-people' reform, accentuated by the reaction to the 2011–2012 'For Fair Elections' movement and the annexation of the Crimea in 2014. The present discourse of the Putin administration emphasizes a concern for the people and a rejection of the neoliberal reforms of the 1990s. One can argue that under Putin, a 'soviet-style neoliberalism' (Hemment, 2009) has developed, a form of neoliberal politics coupled with populist and nationalist values and in ostensible opposition to the economic reforms of the 1990s that traumatized most of the population. Indeed, Putin uses populist rhetoric and turns back to the 'hard-working' and 'conscientious' 'ordinary folk', acknowledging the demand for a socially progressive state. He addresses his speeches to 'ordinary citizens' and 'people who work' and 'love Russia'.[3] His incontestable popularity is largely rooted in the fact that he gives ordinary people something resembling social and political recognition.

The liberal and democratic opposition, for its part, continues to use the rhetoric of the 1990s. The liberal Moscow intelligentsia considers 'mass post-Soviet people', especially poor people from the regions, as 'paternalistically minded', authoritarian, cynical, and materialistic (Gudkov, 2012; Gudkov, Dubin, & Zorkaâ, 2008). This contemptuous discourse persisted during the 'For Fair Elections' movement, which explains the weak attraction that democratic protests in Moscow exerted on 'ordinary people' from the regions (Judah, 2013). In a way, the rhetoric of the liberal opposition reproduces the populist discourse by inverting it (Yudin, 2017): its anti-populist rhetoric supports the myth of the enlightened minority and the unenlightened majority.

In contrast, the patriotic and populist discourse of the Kremlin, regardless of its instrumentalist, manipulative, and demagogic aspects, provides some cues for ordinary people to recover a kind of social consciousness and a sense of social inequality. Surveys confirm a growing sense of social cleavage and inequality (Levada Centre, 2016; Mareeva & Tikhonova, 2016; VTsIOM, 2016), findings also made in our ongoing research on everyday nationalism in Russia. People have become better at figuring out the society they live in; they identify deep social, rather than national or ethnic, cleavages. In-depth interviews with people, from different cities and of different socio-demographic profiles, about their everyday lives show a high proportion of them denouncing the rising social inequality. They advocate for workers, those who produce (often considered 'the genuine patriots'), for pensioners 'who have worked all their life for this country', and for the countryside and the regions that face social breakdown. They point out cleavages between the rich and the working people, between the blue-collar workers and employers or office workers, between the regions and the big central cities. As a slater from Astrakhan noted in June 2016 about the economic and political elites, 'Everything has been seized, it's business, it's profitable for them, do you understand, it's profitable to take everything from the workers, to pay them so little'. Or a 60-year-old Moscow female factory economist (February 2017) who asked: 'How it is possible that one person earn 5,000 roubles and the other 5 million, if they work at the same enterprise!'

To a certain extent, the rising popularity of opposition leader Alexei Navalny, and the mass participation in the protest days against corruption that he initiated in March and June 2017 demonstrate the growing strength of socio-economic motives to protest. Interviews conducted by colleagues from the Laboratory of Public Sociology (ps-lab.ru) and my own research, as well as videos from demonstrations in regional towns, all show that many people took to the streets to voice their discontent with the state of public services, healthcare, public education, culture, or roads. They were pushed to protest by their dissatisfaction with wide social inequalities between a small group of the rich and the majority of the poor, between the prosperous central cities and the neglected and remote regional towns.

These new representations along social cleavages have been made possible, paradoxically, by the patriotic discourse of the power holders and the imaginations that it opens up. Whatever the aims of the ruling elite's patriotic discourse, it gives people the ability of thinking of themselves in terms of a nation or a large common entity that embraces the whole country. It thus constitutes one of the channels through which people can project themselves into wider society and break the tendency towards individual, familial, or local self-entrenchment.

Instead of creating national consensus, the populist and patriotic discourse of the ruling elite produces cleavages and social consciousness. And although the discourse is accepted by the majority, it also raises a sense of critical discernment. In interviews, people have their own understanding of patriotism and often criticize the Kremlin's patriotic project because of the disconnection between words and deeds (claiming to be patriots, but not acting accordingly) and because of the non-social content of the project. The populist discourse conveys general social categories, such as 'the hardworking people', 'the ordinary folk', and 'ordinary citizens', that can produce new social identifications. The feeling of national revival opens the way to look at the society more critically and less consensually, to address grievances and claims on the state. As a market woman in Astrakhan explained (June 2016):

> Putin asks for patriotism, he helps all the people, all, Syrians, Ukrainians, all! Why doesn't he help us? I am a nurse and have to stand here to feed my family. How can I survive with such a low wage? 6,000 rubles! Does Putin could survive with such a wage? Why doesn't he help the nurses, people who work hard?

It seems that among the various meanings given to patriotism from below, the most widespread can be described as social critical patriotism. People claim that patriotism should involve respect for 'the people', not the establishment, the oligarchs, or the rich. Here, 'the people' does not refer to all people, but rather to those who work for Russia and participate in its economic development and welfare. Among the most widespread claims on the state on behalf of these genuine patriots are demands for higher wages, more recognition, cheaper housing, and public services.

As a whole, it seems that the rise of patriotic sentiment and the populist discourse conveys politicization and subjectivation, albeit within certain limits. Social critical patriotism and social consciousness do not necessarily mean the capacity to mobilize and overt dissent, although there is a rise in grassroots initiatives, labour protests, and anti-corruption mobilizations all over the country.

Conclusion

Russia is a perfect illustration of the destructive consequences of widespread precarization and social insecurity on agency, social consciousness, and the relation to politics. Desubjectivation, atomization, and depoliticization were central processes that developed in Russian society after the fall of the communist regime. Resistance to neoliberal capitalism and the rise of social inequalities has always been weak. This does not mean that people were apathetic and inactive; they just did not raise these questions. Grassroots social movements and labour protests developed all around the country throughout post-Soviet times, particularly since the later 2000s when economic, social, and political stabilization allowed people to recover some sense of stability. These mobilizations were mostly small-scale, local, rooted in daily life, and focused on pragmatic problems.

The scattered character of these mobilizations is due to social disorientation, a weak solidarity and sense of social belonging, widespread feelings of powerlessness, and political disenfranchisement. A reverse tendency towards politicization and solidarization began to show up only recently, in line with the development of a new patriotic and populist discourse. Among factors such as a new wave of economic hardship experienced by the majority of the population, which interrupted nearly a decade of relative well-being, the new discourse of the Kremlin seems to provide a framework for thinking in some general and political terms and for recognizing some outrageous features of current Russian society, social inequality being the foremost.

Outrage and anger about inequality are not frequently expressed publicly, although they feature as common issues in everyday conversations. However, these feelings, if commonly shared, can fuel and sustain protest. In whatever way, they provide the ground on which a social critical, which is one branch of a populism from below, can develop. There is indeed some evidence that antagonistic – or agonistic, in Mouffe's (2005) terms – views are emerging regarding what constitutes 'the people' and what is good for them. Thus, there are signs of developing forms of politicization and social imaginary that could, in time, give way to mass social justice movements.

Notes

1. Everyday nationalism is being investigated as part of ongoing research entitled 'Patriotism in Contemporary Russia', under the supervision of the author and supported by the grant for the Foundation for Support of Liberal Education at the National Research University Higher School of Economics St. Petersburg.
2. Data from the Institute of Collective Action, Moscow (IKD).
3. Vladimir Putin, speech held on 23 February 2012, at Luzhniki stadium in Moscow, at a rally in opposition to the 'For Fair Elections' movement.

Acknowledgements

The author wants to express her gratefulness to the special issue's editors and to all the anonymous reviewers who contributed to the enhancement of the paper.

Disclosure statement

No potential conflict of interest was reported by the author.

ORCID

Karine Clément ⓘ http://orcid.org/0000-0002-2710-928X

References

Aidukaite, J., & Fröhlich, C. (2015). Struggle over public space: Grassroots movements in Moscow and Vilnius. *International Journal of Sociology and Social Policy, 35*(7/8), 565–580.

Angelovici, M., Dufour, P., & Nez, H. (2016). *Street politics in the age of austerity: From the indignados to occupy.* Amsterdam: Amsterdam University Press.

Ashwin, S. (1998). Endless patience: Explaining Soviet and post-Soviet social stability. *Communist and Post-Communist Studies, 31*(2), 187–198.

Bikbov, A. (2012). The methodology of studying 'spontaneous' street activism (Russian protest and street camps, December 2011–July 2012). *Labouratorium, 2,* 130–163.

Biziukov, P. (2017). *Dynamics of labour protest in Russia.* Manuscript submitted for publication.

Boltanski, L., & Chiapello, E. (1999). *Le nouvel esprit du capitalisme.* Paris: Gallimard.

Breman, J., & van der Linden, M. (2014). Informalizing the economy: The return of the social question at a global level. *Development and Change, 45,* 920–940.

Burawoy, M. (1999). *The great involution: Russia's response to the market.* Unpublished manuscript. Retrieved from http://burawoy.berkeley.edu/Russia/involution.pdf

Clément, K. (1999). *Les ouvriers russes dans la tourmente du marché.* Paris: Syllepse.

Clément, K. (2004). Formal'nye i neformal'nye pravila: Kakov optimum? [Formal and informal rules: What balance?]. In V. Yadov, J. De Bardeleben, & S. Klimova (Eds.), *Stanovlenie trudovyh otnošenij v postsovetskoj Rossii* [The making of labour relations in post-Soviet Russia] (pp. 135–192). Moscow: Akademičeskij Proekt.

Clément, K. (2008). Russie: Récent renouveau des luttes du travail et du mouvement syndical. In J. Magniadas & R. Mouriaux (Eds.), *Le syndicalisme au défi du 21e siècle* (pp. 198–235). Paris: Syllepse.

Clément, K. (2013a). *Gorodskie dviženiâ Rossii v 2009–2012 godah: Na puti k političeskomu* [Urban movements in Russia, 2009–2012: On the road towards politics]. Moscow: Novoe literaturnoe obozrenie.

Clément, K. (2013b). Civic mobilization in Russia: Protest and daily life. *Books and Ideas* (June 14, 2013). Retrieved from http://www.booksandideas.net/Civic-Mobilization-in-Russia.html

Clément, K. (2015a). Unlikely mobilisations: How ordinary Russian people become involved in collective action. *European Journal of Cultural and Political Sociology, 2*(3–4), 211–240.

Clément, K. (2015b). From 'local' to 'political': The Kaliningrad mass protest movement of 2009–2010 in Russia. In K. Jacobsson (Ed.), *Urban grassroots movements in Central and Eastern Europe* (pp. 163–194). Farnham: Ashgate.

Clément, K., Demidov, A., & Miryasova, O. (2010). *Ot obyvatelej k aktivistam: Zaroždaûŝiesâ social'nye dviženiâ v nynešnej Rossii* [From ordinary people to activists: Emerging social movements in contemporary Russia]. Moscow: Tri kvadrata.

Danilova, E. (2014). Neoliberal hegemony and narratives of 'losers' and 'winners' in post-socialist transformations. *Journal of Narrative Theory, 44*(3), 442–466.

Danilova, E., & Yadov, V. (1997). Social identifications in post-Soviet Russia: Empirical evidence and theoretical explanation. *International Review of Sociology, 7*(2), 319–335.

Daucé, F., Laruelle, M., Le Huérou, A., & Rousselet, K. (2015). Introduction: What does it mean to be a patriot? *Europe-Asia Studies, 67*(1), 1–7.

Della Porta, D. (2015). *Social movements in time of austerity: Bringing capitalism back into protest analysis.* Cambridge: Polity Press.

Erpyleva, S., & Magun, A. (2014). *Politika apolitičnyh. Graždanskie dviženiâ v Rossii 2011–2013 godov* [The politics of the 'apolitical': Citizens movements in Russia, 2012–2013]. Moscow: Novoe literaturnoe obozrenie.

Evans, A. (2015). Civil society and protests in Russia. In C. Ross (Ed.), *Systemic and non-systemic opposition in the Russian Federation: Civil society awakens?* (pp. 15–34). Farnham: Ashgate.

Gabowitsch, M. (2016a). Are copycats subversive? Strategy-31, the Russian runs, the immortal regiment, and the transformative potential of non-hierarchical movements. *Problems of Post-Communism, 62,* 1–18.

Gabowitsch, M. (2016b). *Protest in Putin's Russia.* Cambridge, MA: Polity Press.

Gerbaudo, P. (2017). The indignant citizen: Anti-austerity movements in southern Europe and the anti-oligarchic reclaiming of citizenship. *Social Movement Studies, 16*(1), 36–50.

Gladarev, B., & Lonkila, M. (2013). Justifying civic activism in Russia and Finland. *Journal of Civil Society, 9*(4), 375–390.

Glasius, M., & Pleyers, G. (2013). The global moment of 2011: Democracy, social justice and dignity. *Development and Change, 44*(3), 547–567.

Greene, S. (2014). *Moscow in movement: Power and opposition in Putin's Russia.* Stanford, CA: Stanford University Press.

Gudkov, L. (2012). Social'nyj kapital i ideologičeskie orientacii [Social capital and ideological orientation]. *Pro et Contra, 16*(3), 6–31.

Gudkov, L., Dubin, B., & Zorkaâ, N. (2008). *Postsovetskij čelovek i graždanskoe obŝestvo* [Post-Soviet man and civil society]. Moscow: Moskovskaâ škola političeskih issledovanij.

Hemment, J. (2009). Soviet-style neoliberalism? Nashi, youth voluntarism, and the restructuring of social welfare in Russia. *Problems of Post-Communism, 56*(6), 36–50.

Jacobsson, K. (2015). *Urban grassroots movements in Central and Eastern Europe: Introduction.* Farnham: Ashgate.

Judah, B. (2013). *Fragile empire: How Russia fell in and out of love with Vladimir Putin.* New Haven, CT: Yale University Press.

Koveneva, O. (2011). Les communautés politiques en France et en Russie. *Annales: Histoire, Sciences Sociales, 66*(3), 787–817.

Krivonos, D. (2015). State-managed youth participation in Russia: The national, collective and personal in Nashi activists' narratives. *Anthropology of East Europe Review, 33*(1), 44–58.

Levada Centre. (2016). *Points of disagreement in the society* [Press release in Russian]. Retrieved from http://www.levada.ru/2016/06/27/tochki-raznoglasij-v-obshhestve/

Mandrik, M. (2017). *Social inequality and protest politicization: conservative mobilization under the conditions of social-economic crisis in 2015–2017 Russia* (Master's thesis). European University at St. Petersburg, St. Petersburg.

Mareeva, S., & Tikhonova, N. (2016). Public perceptions of poverty and social inequality in Russia [in Russian]. *Mir Rossii, 25*(2), 37–67. http://www.isras.ru/files/File/publ/Mareeva_Tikhonova_mr16_2_2.pdf

Morris, J. (2016). *Everyday post-socialism: Working-class communities in the Russian margins.* New York, NY: Palgrave Macmillan.

Mouffe, C. (2005). *On the political.* London: Routledge.

Novokmet, F., Piketty, T., & Zucman, G. (2017). *From Soviets to Oligarchs: Inequality and Property in Russia, 1905–2016* (No. w23712). National Bureau of Economic Research. Retrieved from http://www.nber.org/papers/w23712.pdf

Oushakine, S. (2000). In the state of post-Soviet aphasia: Symbolic development in contemporary Russia. *Europe-Asia Studies, 52*(6), 991–1016.

Robertson, G. B. (2010). *The politics of protest in hybrid regimes: Managing dissent in post-communist Russia.* Cambridge: Cambridge University Press.

Scott, J. (1985). *Weapons of the weak: Everyday forms of peasant resistance.* New Haven, CT: Yale University Press.

Sennett, R. (1998). *The corrosion of character: The personal consequences of work in the new capitalism.* New York, NY: Norton.

Smyth, R., Sobolev, A., & Soboleva, I. (2013). Patterns of discontent: Identifying the participant core in Russian post-election protest. Paper presented at the Russia's Winter of Discontent: Taking Stock of Changing State-Society Relationships Conference, September 6–7, Uppsala University, Sweden. Retrieved from http://www.ucrs.uu.se/digitalAssets/172/a_172857-f_discontentconfsmythpatternsofdiscontent.pdf

Standing, G. (2011). *The precariat: The new dangerous class.* London: Bloomsbury Academic.

Tykanova, E. (2012). *Contested urban spaces: Russia and Europe in comparative perspective.* Working Paper WP 2012–09, Centre for German and European Studies. Retrieved from http://www.zdes.spbu.ru/assets/files/wp/2012/WP_2012_9 Tykanova.pdf

Volkov, D. (2012). The protesters and the public. *Journal of Democracy, 23*(3), 55–62.

VTsIOM (All-Russian Centre for the Study of Public Opinion). (2016, November 15). *Together or apart? Social contradictions in modern Russia* [Press release in Russian]. Retrieved from http://wciom.ru/index.php?id=236&uid=115947

Yudin, G. (2017, March 23). 'Scratch a Russian liberal and you'll find an educated conservative': An interview with Russian sociologist Greg Yudin. *LeftEast.* Retrieved from http://www.criticatac.ro/lefteast/scratch-a-russian-liberal-and-youll-find-an-educated-conservative-an-interview-with-sociologist-greg-yudin/

Zhuravlev, O., Savelyeva, N., & Erpyleva, S. (2014, October 6). Apoliticism and solidarity: Local activism in Russia. *LeftEast.* Retrieved from http://www.criticatac.ro/lefteast/apoliticism-and-solidarity-local-activism-in-russia/

Mapping movement landscapes in South Africa

Karl von Holdt and Prishani Naidoo

ABSTRACT
The concept of a movement landscape is used to analyse continuities and changes in popular mobilization since the end of formal apartheid. Focusing on four different episodes of protest since 1997, the article examines their relationship to the ANC movement and traditions, and their organizational forms. It finds a general theme of fluid and ephemeral organization, and a distrust of formal hierarchal organization, that is relatively new in South Africa. The Marikana strikes produced the most far-reaching organizational realignments, while the student struggles generated the most innovative re-imaginings of political forms and discourses. It concludes that although there have been critiques of and challenges to the ANC tradition, and experiments with new forms of organizing, they have not produced alternatives that have lasted or dislodged the dominant approaches defined and popularized by the ANC movement.

Two decades of electoral democracy in South Africa have been accompanied by cycles of contention that have deployed both collective action and strategic litigation based on the new regime of constitutional rights created in the negotiated transition to democracy (1991–1996). The intensity of contention and popular mobilization has led some to consider South Africa the 'protest capital of the world' (Alexander, 2010, 2012a); and events such as the Marikana massacre and the subsequent strike wave (2012–2014), and the student movements of 2015–2016 have been discussed as 'turning points' in popular struggle (Alexander, 2013). Yet, none of these events has produced a decisive rupture with the dominant politics of the Congress Alliance led by the African National Congress (ANC), despite a steady decline in electoral support for the ANC over these decades (Veriava, 2015).

With this article, we explore four distinct forms or episodes of popular mobilization, in an effort to surface both the reproduction of older organizational forms and practices and the partial emergence of what appear to be new repertoires and organizational forms which may portend some kind of reorientation of popular struggle. This is a partial selection determined by familiarity – our own research and that of close colleagues, as well as activist participation – with a view to comparing a variety of different kinds of movements, periods, and sites of struggle. In this article, we do not aim to scan the entire range of movements active in South Africa, nor do we engage in a comprehensive review of the literature. Rather, we seek to drill down into four promising case studies in a search for common features and new departures.

The concept we develop in order to think through the problem is that of a movement landscape (Cox, 2016a, 2016b).[1] A movement landscape constitutes the terrain on which popular mobilization and movements have to make themselves visible and position themselves in order to pursue their goals. We can identify two elements of a movement landscape. The movement landscape is shaped by successive histories of popular struggle, which crystallize in particular organizational structures, and lay down patterns of grievances or claims, repertoires of mobilization and action, and the symbolic forms and languages, as well as the silences, which tend to define what may or may not be articulated and heard. Such a landscape is shaped not only from below, but also by institutions, authorities, and elites from above – regimes of citizenship, institutions of democracy, rights that may be enshrined in constitutions and laws, promises and policies, systems of certification, recognition and regulation that legitimate certain organizations and claims, and derecognize or delegitimate others. Importantly, movement landscapes reflect ongoing contestation over what is considered to be political and about what is thought to be possible through political struggle. Popular forms of political engagement may push up against these limits, or transgress them, redefining the terms of engagement between elites and subalterns and substantially altering the movement landscape. They may, in other words, reproduce prevailing terms of incorporation, negotiate an alteration to them, transgress them profoundly, or even on occasion produce a rupture that substantially alters political relations. The concept of movement landscape is intended to capture therefore the durability of the current order of things as well as the change – erosion, fracture, the sedimentation of new projects and practices, and the ruptures of formative events which may dramatically tear apart and reconfigure landscapes (Roseberry, 1994).

The concept of a movement landscape is intended to provide a perspective that overcomes some of the limitations of the dominant 'contentious politics' school of social movement analysis noted by several scholars (for example, Barker, 2013; Barker et al., 2013; Burawoy, 2017; Gabowitsch, 2017; Runciman, 2017; Zibechi, 2012), and draws attention to history, durability, and change, locates a movement in relation to the variety of different movements, and assumes a fluidity of networks and porosity of boundaries among movements as well as between movements, elites, and states.

We argue that the negotiated transition in South Africa (1990–1996) constituted a formative event as envisaged above, through which the black majority were incorporated as citizens, a new constitution elaborating the rule of law was drafted, and the ANC was enshrined as the dominant political force both in the institutional structures of the new order, as well as informally through deeply rooted networks across society. The ANC is deeply implicated in the founding of the new order, this position cemented not only by its historical role in achieving liberation and negotiating a new order, but also by a wide range of government policies and patronage systems, securing a variety of pathways for elite formation and enrichment, as well as a degree of redistribution to subalterns. Celebrated as 'the democratic breakthrough' by ANC leaders, over time few would remember this founding moment as the outcome of a negotiated settlement, a set of compromises, rather than a straightforward triumph.

Our working argument is that the constitutive moment of democratic transition laid down the basic contours of the movement landscape, in the form of both new rights and institutions, as well as the deep histories of popular struggle through which the transition to democracy was produced. Thus, the new democratic institutions of citizenship, participation, and incorporation dramatically restructured the apartheid movement landscape, incorporating movement constituencies in unprecedented ways and establishing new spaces and constraints. This does not mean that the history of struggle is irrelevant to the movement landscape. On the contrary, the organizational forms, repertoires, and symbolic power that were mobilized in the ANC tradition over some 80 years of struggle, but most importantly during the 1980s, continue to define and shape popular struggles today with paradoxical effects – reproducing allegiance to the ANC while providing memories, repertoires, and claims that may be mobilized against it (Hart, 2013).

Analogous founding moments characterize each of the BRICS countries, whether these emerged from decolonization, Communist revolution, or democratic transitions from dictatorship (see Cox, 2016a, for the Irish case as a postcolonial society).

The question this article seeks to answer, therefore, is to what extent post-apartheid struggles and movements are contained within and reproduce the movement landscape and the relationships embedded within it, and, conversely, to what extent they transgress such limits. A large part of the movement landscape organized around the ANC constellation of organizations (or 'Congress')[2] is deeply shaped by the history of the national liberation movement. More informally, ANC networks ramify through society, influential in professional associations, religious organizations, black business forums, and so on. This is the organizational terrain on which – and against which – popular mobilizations and protests take shape, thus providing a set of organizations, repertoires, symbolic powers, and landmarks – a set of understandings which are mutually intelligible by activists, communities, elites, and the state – but also a set of constraints, limitations, and silences which it is difficult to transgress. Indeed, transgression may carry high costs.

Our examination of four different sets of popular mobilization in this article proceeds along two axes – relation with the Congress constellation of organizations, and forms of organizing – as these provide strong indicators for assessing the question we have set ourselves: whether any of these mobilizations begin to break the organic relationship between popular mobilization and Congress politics laid down in the movement landscape. The first axis of enquiry in turn opens up two questions – first, the question of the organizational relationship with the Congress constellation and, second, the relationship with Congress traditions of mobilization. The four mobilizations we discuss include:

- The 'new social movements'[3] that emerged in the late 1990s and early 2000s,
- the localized community protests which started to become something of a wave from the mid-2000s,
- the Marikana massacre in 2012 and the strikes and organizational shockwaves this produced,
- and the mass movement of students and workers that mobilized at universities in 2015/2016.

We examine each of these mobilizations in sequence, precisely in order to understand whether there is any continuity or resonance between them, and whether or not new repertoires, organizational forms, or landmarks are being laid down in the movement.

1. Contesting the terms of democracy – 'new social movements'

The idea that 1994 was a 'democratic breakthrough' was limited by the ascent of neoliberal policies and logics globally. In spite of fierce contestation from within the party and the broader alliance, in 1996, the ANC government adopted a neoliberal macro-economic policy framework that would constrain the possibilities for change (Bond, 2000; Marais, 2001). With the framework declared 'non-negotiable' by their leaders, members of the Congress movement who continued to be critical of its adoption in public spaces outside of the movement were labelled 'ill-disciplined' and heavily censured, most often expelled. By the end of the 1990s, many of these (and other disgruntled) activists found allies amongst different groups of people brought together by their common experience of the various negative effects of this policy turn – township residents unable to pay for basic services demanding an end to their being cut off from water and electricity and evicted from their homes as the logics of commodification and privatization were enforced; people living and working with

HIV/AIDS unable to afford the cost of anti-retrovirals and care more generally as state spending on health was cut; workers retrenched as a result of trade liberalization policies and privatization; workers in contract, seasonal, part-time, casual, and outsourced jobs as the regime of flexible labour took hold; people affected in different ways by the 'willing buyer, willing seller' approach to land redistribution.

Between 1997 and 2006, then, the movement landscape was dotted with eruptions of protests that over time provided the conditions for the emergence of more formal and structured organizations through which ongoing and sustained action and intervention could be facilitated. The Anti-Privatization Forum (APF), the Concerned Citizens' Forum (CCF), and the Anti-Eviction Campaign (AEC) were established in Johannesburg, Durban, and Cape Town, respectively, in struggles for free basic services. At a national level, the Treatment Action Campaign (TAC) emerged to fight for free anti-retrovirals and other necessary resources for people living with HIV/AIDS, and the Landless People's Movement (LPM) was formed to co-ordinate struggles related to land and farm work.

Together representing the first set of movements to emerge outside of the Congress tradition after 1994, they were referred to as the 'new social movements' (Ballard, Habib, & Valodia, 2006; McKinley & Naidoo, 2004; Naidoo & Veriava, 2005). Although these movements might have been critical of the ANC government and leadership of the alliance for facilitating these policy shifts, a majority of their members came from Congress structures. This often saw the framing of demands by these movements around the claim that the ANC government was not fulfilling the mandate it inherited from the liberation movement by moving away from commitments made during the struggle against apartheid. It also meant that their political repertoires and organizational forms and cultures often resembled or drew from those of Congress formations (Naidoo & Veriava, 2005; Runciman 2012, 2015).

But these movements were also usually open spaces to which anyone could belong regardless of their political affiliation. This meant, then, that there were debates within these movements about political orientation (in particular to the different formations making up the ANC alliance) as members from Black Consciousness, Pan Africanist, and various socialist and Marxist traditions came into conversation with ANC members as well as individuals who chose not to align to any formal political party or tradition. Depending on the relationship of its founding members and leaders to the ANC and its alliance partners, and the nature of the issues being taken up in its struggles, each movement adopted its own modes of critical engagement with the ANC and its aligned formations (Dwyer, 2004; Naidoo & Veriava, 2005).

Across this set of movements, the immediacy of the problems at the heart of the protests produced approaches that combined both legal and illegal forms of action, demonstrating the simultaneous turn to the law and other processes promising redress through state institutions and policies, and the turn to past practices employed in the struggle against apartheid, such as the illegal reconnection of houses to water and electricity. In the struggles against neoliberal approaches, this repertoire grew to include the destruction of prepaid water meters as well as the illegal bypassing of prepaid electricity meters, and campaigns of defiance that openly disobeyed laws protecting the interests of transnational capital over those of poor people across the globe. In the collectivization and politicization of such acts, movements provided a counter discourse and logic to that of privatization, commodification, and trade liberalization, drawing boldly on commitments made in the liberation movement to free access to the resources necessary for a good life (Desai, 2002; Friedman & Mottiar, 2004; Naidoo & Veriava, 2005; Van Heusden & Pointer, 2005).

The main differences across this set of movements (also often reflected within movements) related to organizational forms, and contestation of national and local government elections. Both speak to conflicting understandings and imaginings of politics and possibilities for political space. Disagreements over the value of hierarchical versus non-hierarchical structures, and the need for leaders versus more collective and participatory forms of decision-making, tended to dominate. Ultimately, the shape and form that a collective took changed over time, and was influenced by past organizational experiences of members, the character of the issues and engagements it was involved in, and the dominant political traditions and ideological dispositions of its members. These debates were also influenced by practices in the alter-globalization movement, a coming together of groups across the world challenging the adoption of neoliberal policies.

Notwithstanding differences, by 2001, as the fight against neoliberal policies grew globally, it was clear that in South Africa too, there was a set of movements questioning and acting against the neoliberal policy reforms of the ANC government. This was perhaps most starkly evident in 2001 and 2002 when over 20,000 people marched during the World Conference Against Racism and the World Summit on Sustainable Development, respectively, both United Nations conferences hosted by the South African government. Both times, the aim was to highlight and demand an end to the adoption of neoliberal policies, which, in South Africa, they argued, perpetuated racism and prevented sustainable development. For the first time since 1994, the movement landscape was coloured by a collective force outside of the ANC that saw the coming together of very different groups in a common stand against neoliberal policies (Desai, 2002; Naidoo, 2002; Naidoo & Veriava, 2005).

But by 2006, most of these movements were in decline or no longer in existence. Increasing state repression took its toll, and the state's incorporation of certain of their demands immobilized some. Other factors included power struggles between different political factions, and the poor handling of conflict over race, class, and gender differences, as well as over control of increasing amounts of donor money. Nevertheless, many of the affiliate community formations that made up these movements survived at a local level (McKinley, 2012; Naidoo & Veriava, 2013). Some, influenced by the new legal and democratic institutions in the landscape, over time came to resemble NGOs more than social movements. Although the impact of these movements continues to be felt in the lives of the poor and vulnerable, their fragility in sustaining co-ordinated action across a range of local level formations was ultimately revealed in their decline.

2. Community protests

A pattern of community protests started emerging outside the structures of formal social movements such as the APF around 2004, though in many instances, they were raising issues similar to those raised by the preceding social movements. Isolated protests of this kind had taken place since the 1990s, but from 2005 there was a steady increase, with a sharp spike to a new and sustained level of protests in 2009 (Alexander, 2012a; Municipal IQ, 2012, 2017; Runciman et al., 2016).

In most instances, protests have been directed against local municipal authorities governed by the ANC and mobilized over grievances including housing, water and other municipal services, lack of jobs, corruption over the allocation of houses and plots, or the disappearance of municipal funds and lack of responsiveness from authorities. Typically, protests take the form of a series of marches, the presentation of memoranda of grievances, and mass community meetings, and may escalate to include work stayaways, street barricades, running battles with the police, and burning of municipal buildings or municipal councillors' homes (Alexander, 2010, 2012b; Langa & von Holdt, 2012; Von Holdt, 2014; Von Holdt et al., 2011).

A study of the internal life of the networks and crowds that emerge during these episodes of protest reveals a complex amalgam of relatively autonomous mobilization and action from below together with elite politics from above (Langa & von Holdt, 2012; Von Holdt et al., 2011). The study found a variety of organizational forms – for example, the ANC Youth League, the SACP, the South African National Civic Organization, and residents' committees – all sharing a deep connection with ANC networks and structures. Grievances were real and the popular protests had mass support and were characterized by popular initiative. Yet, paradoxically, protest participants were often dismissive of protest leaders, commenting that they were pursuing their own interests, specifically electoral office in the town council or preferential access to council resources and tenders. The ANC councillors who were targets of the protests made the same allegations. Notwithstanding this cynicism, thousands of residents participated in protests and marches. And in the aftermath of the protests, protest leaders were indeed frequently incorporated in the local council or municipal administration, provided with access to tenders or deployed into positions in the ANC constellation with minor benefits. The protest movements themselves reveal – and are shaped by – a fierce intra-elite struggle over processes of elite formation through access to lucrative jobs, resources, and patronage networks.[4]

The masses of residents are not passive victims of this process, though. They demonstrate their own ability to put pressure on the leaders, call them to account, and initiate actions such as stayaways, marches, and street battles autonomously. Recognizing that the ANC is the dominant organization in local institutions and in the community, they make use of elite leaders in order to be recognized and heard by powerholders within the ANC, just as much as the leaders position themselves in protests to establish a constituency and a power base through which to challenge incumbents and reconfigure relations within the local structures. Elements of this tension are discernible in the following quote from a protest leader:

> We could see that what the community was fighting for was genuine, and that as members of the ANC Youth League we were quiet about this thing. Why can't we tap into this thing and start channelling this thing to right directions within us here as the ANC? We did that.

The protest movements thus have a *dual character* (Langa & von Holdt, 2012; for similar arguments see Dawson, 2014, 2017; and Mukwedeya & Ndlovu, 2017), and it is these tensions between mobilization from below and positioning from above that shape the trajectories of protest.

The organizational picture that emerges in the aftermath of protests is of the durability of the formal structures of the ANC constellation and the relatively ephemeral nature of autonomous community-based structures. None of these cases produced enduring community-based organizations or movements whose primary focus is on representing, organizing, and mobilizing residents. What they did leave, however, were a legacy of organizational forms and organizational repertoires – ranging from formal ANC-linked structures to informal and nominally independent ones – which could once again be appropriated for popular mobilization when the need arose.

To sum up, then, many community protests take place through structures and activist networks located within the ANC constellation of organizations. Both leaders and subalterns situate their struggles on the terrain of the ANC, deploying recognizable repertoires, symbols, and discourses of struggle, appearing to reproduce the forms and landmarks of the movement landscape. Nonetheless, these dynamics of incorporation are inflected by tensions and new repertoires as well. The localized community protest movement in South Africa has forged an innovation in the movement landscape – that is, the appropriation of ANC networks and structures to mobilize protest, and with increasing frequency, violent protest, against other ANC networks and structures, partly in the service of factional struggles, but also in the service of raising the grievances and demands of communities. By so doing it destabilizes the ANC, legitimates popular agency, and raises grievances and demands that elites may prefer to be silenced. This is an enduring and expanding set of repertoires, as evidenced by the escalation in community protests over the past decade.

It is important to highlight the moment of autonomy that is present as a tension in many such protests – the networks and actions from below that put pressure on protest leaders and escalate the momentum of popular action – and the fact that some among the protest leadership are more responsive to these dynamics than others who are focused on their own trajectories towards power. These aspects suggest the potential that is present in localized community protest for a break with the ANC as the terrain of protest, and the ever-present possibility of new organizational forms and repertoires which present alternatives to the ANC.

3. Marikana and the remaking of the trade union movement

In August 2012, the South African police shot and killed 34 striking mine workers at Marikana, part of the platinum belt which stretches across the northern provinces of South Africa and is the site of multiple social transformations driven by the massive expansion of the platinum mining industry. The Marikana massacre occurred at the midpoint of a wave of mass strikes that in turn shook each of the three biggest platinum mining companies in the world during that year. The following account draws from an extensive literature that has already been published on these events (including Alexander, Sinwell, Lekgowa, Mmope, & Xezwi, 2013; Chinguno, 2013, 2015a, 2015b; Sinwell, 2013, 2015; Sinwell & Mbatha, 2016).

In all three cases, the strikes were led by independent worker committees, bypassing or directly attacking the workplace structures of the National Union of Mineworkers (NUM), which since the 1980s had been the dominant mining union and the biggest affiliate of COSATU. Violent conflict between strikers and NUM representatives escalated, and eventually, the union collapsed across most of the platinum belt, to be replaced by a small breakaway, the Amalgamated Mining and Construction Workers Union (AMCU).

Why did the organizational structures of the NUM collapse so dramatically across the three mining companies? Chinguno (2015a) provides us with a detailed critique of the way the union shaft stewards, officials, and structures had been incorporated into management structures and practices, with shaft stewards and office bearers benefiting from access to improved salaries, advanced training, the opportunity to dispense patronage in the form of jobs and accommodation, and clear promotion paths into management. Deepened institutionalization led to 'class capture' and growing anger towards the union for its failure to respond to worker issues (see also Sinwell, 2015, pp. 597–599).

Thus the strikes took the form of wildcat 'unprotected' actions, organized outside of and against the procedures and institutions of the industrial relations system that had been established post-apartheid, and that in turn was based on many of the procedures, forums, and rights that had been fought for and established in practice by the union movement in the anti-apartheid struggles of the 1980s. Furthermore, the demands they made breached the tacit understandings of long-established negotiating practices.

But the demands represented a symbolic rejection of the established collective-bargaining system and its wage increases that tend to oscillate around a figure just above inflation – and a claim for a much greater share of the wealth generated by mining. Management responded with mass dismissals, refusal to negotiate, and the deployment of security and police. Ultimately, the strikes led to a rupture with the Tripartite Alliance and the ANC, though it did not start out that way. Workers were rejecting their union, not the ANC. However, the NUM was an important organization in the

Alliance, the biggest union in COSATU, and one of the staunchest supporters of the ANC in the federation. In the end, COSATU, the ANC, and the SA Communist Party all swung behind the NUM, castigating the workers variously as anarchists, lumpen proletarians, and party to a management plot or to a 'third force' attack on the ANC. After the Marikana massacre, government, the ANC, and the Communist Party came out in support of the police action against the strikers and the ANC. It was clear that the strike was regarded as transgressive, provoking multiple mechanisms of delegitimation and justifying the use of extreme state violence.

By 2013, all three of the companies were negotiating recognition agreements with AMCU, which had displaced the NUM as the majority union. By 2014, AMCU was the dominant union across the platinum belt. Committee members became shaft stewards and office bearers in the new union's branches. Nonetheless, there were continuing tensions between committee members and the union, particularly its national leadership, as they accused it of imposing top-down decisions and operating in a way that was analogous to the NUM. In at least two of the companies – Impala and Anglo Platinum – elements of the worker committees continue to meet autonomously from the union in order to preserve a degree of worker independence.

The attack on central institutions of the post-apartheid industrial relations order and the formation of independent worker committees was clearly a break with key features of the movement landscape established in the era of democracy, though it had roots in the apartheid period. On the platinum belt, independent committees had organized under extreme repression and were pre-cursors to the introduction of the NUM, and in other unions had frequently been organized in an attempt to maintain control over shop stewards from below (von Holdt 2003). While these innovations in the 2012 strikes were reabsorbed into the industrial relations system with the recognition of AMCU, the 2012 events had broader repercussions, ultimately producing new organizations outside the ANC constellation and severely weakening it in the labour field.

Thus, the Marikana strikes inaugurated a series of ruptures in the movement landscape, throwing into question key industrial relations institutions, reducing the NUM, and weakening COSATU and the Congress constellation, producing a new labour federation and providing momentum for a small but robust left-wing political challenge to ANC domination in the form of the Economic Freedom Fighters (EFF). Yet, these organizational initiatives continue to reproduce important elements of ANC traditions, deploying these against the ANC constellation. The new labour Federation, dominated by former COSATU affiliates, is led by a former COSATU general secretary, and its launch was marked by the singing of COSATU and ANC songs. The EFF claims to represent the 'real' ANC and its vision, enshrined in the Freedom Charter. The Marikana strikes came to constitute a transgressive struggle, an assertion that the established 'order of things', and specifically its distribution of wealth and poverty, was no longer acceptable. Workers drew on subaltern understandings, rather than confining themselves to the standards of 'reason' embodied in the institutions and organizations of the official landscape. These constitute an active legacy in the evolving movement landscape.

4. From hashtags to movements?

The newest additions to the movement landscape are student collectives that have grown from mobilizations popularized through hashtags – #TransformWits, #RhodesMustFall, #Outsourcing-MustFall, #FeesMustFall – which have been analysed in vivid accounts by student activists themselves (Chinguno et al., 2017; Langa, 2017). Led by a new generation of mainly students and younger workers, these protests have been significant for how they have mobilized old repertoires from across the political spectrum in a very new context and through very new forms of media.

They have made institutions of higher learning 'sites of struggle'[5] once again, and returned society to battles lost in the early 2000s, including struggles against outsourcing and for free education,[6] as universities started to reorganize along neoliberal lines. Social media facilitated the mobilization of students who would not ordinarily have joined a protest led by the Student Representative Council, and the simplicity of the slogan 'Fees Must Fall' appealed to a majority of students across all kinds of differences, without any particular political or ideological framing (for student accounts, see Malabela, 2017; Mashibini, 2017; Ndlovu, 2017). At the same time, however, songs, slogans, and icons usually associated with one particular political tradition were appropriated by all and made common to the immediate struggle. For example, the name and history of Solomon Mahlangu, an ANC guerrilla who was executed by the apartheid regime, became so defining of those who gathered in struggle in Senate House at Wits that it was the name they chose to give to the building as part of the process of decolonizing the university (Mthombeni, 2017). And, although the protests may have been started by students involved in formal political organizations, the majority of those who joined and sustained the actions were first-timers to protest.

In these new mobilizations, the dominance of the Congress movement has been questioned as many students turned to Black Consciousness, Pan-Africanism, and Black thought more generally in their grappling with the problems identified with the ANC government. Students called for the 'decolonization' of higher education, in particular of the curriculum and institutional cultures. Students identifying as feminist and queer critiqued student organizations for reproducing masculinist, sexist, misogynistic, racist, homophobic, and other exclusionary and prejudicial forms of engagement and organization (Dlakavu, 2017; Jacobs, 2017).

What became the largest of these movements, taking on a national character, began on 13 October 2015, when students at Wits in Johannesburg embarked on protests against a proposed fee increase for 2016 behind the hashtag, #FeesMustFall. Led by elected student leaders who felt let down by the representative structures of governance and decision-making of the university, and joined by outsourced workers engaged in a struggle for inclusion, the protests swelled to unprecedented levels, both at Wits and nationally (Kgoroba, 2017; Mabasa, 2017).

Although the initial protest was led by the Wits Students Representative Council dominated by Congress-aligned formations, it soon overran its origins. In the mass open forums in Solomon House, different political and ideological traditions as well as lived experience and exposure to forms of collective action and protest from the alter-globalization and Occupy movements, produced rich discussions and debates about 'decolonization', the plight of 'the Black child', and calls for more democratic and less hierarchical forms of organization and decision-making. The presence and contribution of feminists amongst the Black women involved have meant the constant questioning and calling out of patriarchy in its various manifestations within the movements. 'Intersectionality' has become part of these discussions as a counter to traditional class analyses that dominate in older movements.

But there were always attempts to control the space by different party political factions, and what might have been productive disagreements often ended in defensive battles between small groups within the common space (Mashibini, 2017). Nevertheless, students were able to unite across their differences, and come together with workers, in mass marches on national government, and occupations and other forms of action at individual institutions, that produced a final agreement to no fee increases in 2016 and a commitment at some institutions (in addition to Wits) to work towards the end of outsourcing. By then, students had deepened their demands to call for free education. But as the ANC and government began to take a firm stand against protesters, a split developed between Congress-aligned student formations, which argued that free education was a long-term demand and that students should return to class and write exams, and a group of students that grew smaller and smaller over time who kept Solomon House occupied until the holiday period (Ndlovu, 2017). The following year was marked by an escalation of violence on campuses as protesting students were met with brute force by private security and the police, and students turned to violence in response. A significantly new feature of the movement landscape as a result of #FeesMustFall is the highly securitized character that has come to define the terrain of struggles in higher education.

Today #FeesMustFall is fractured. Party politics has begun to dominate again, and the high levels of securitization on campuses have disabled even the most radical groups of activists. Nevertheless, the movement landscape has been profoundly affected by the emergence of a new generation of student activists and potent new slogans and discourses, such as decolonization, Black Feminism, intersectionality, and radical democracy. It has also successfully brought together in common struggle previously separate groups, such as workers (albeit limited to campus workers) and students, opening up possibilities for the shaping of collective political subjectivities across class divisions, and the production of new political forms and strategies that do not confine themselves to a singular imagination of 'the revolutionary subject' as 'the worker'. #FeesMustFall holds the potential for a different kind of politics, if those who have come together in its name are able to overcome their factionalism and engage more productively with each other in shaping and taking up struggles. In this way, they could reopen possibilities that were long ago foreclosed by the political cultures inherited from the ANC alliance and other parts of the liberation movement and the choices made by the ANC government in relation both to politics and the economy.

5. Concluding discussion

In this final section, we try to draw out some of the patterns that emerge across the four distinct mobilizations, and their implications for the movement landscape. It is important to note that while the movement landscape as a concept points towards resonances and connections between movements, it also allows us to specify the contrasting features of different regions of the landscape. Thus, while it insists that we include both labour movements and community movements in the same analysis, it also points to the distinct features marking their different regions in the landscape. Labour movements mobilize in the most institutionalized, routinized, and regulated region in the landscape, while community movements emerge in a region with limited organization and regulation. Labour movements mobilize employed workers, and their counterparts are employers; community movements mobilize heterogenous constituencies, many unemployed, in relation to diverse authorities, though most usually the local government. The contrasting features of these two sites determine very different organizational forms, repertoires, and trajectories (see Alexander & Pfaff, 2013), accounting for some of the differences discussed below.

It is quite striking that each of the four mobilizations discussed in this article was originally rooted in some way or another in the Congress constellation of organizations, but from these beginnings pursued varied trajectories which positioned them at a greater or lesser distance from this constellation. The community mobilizations tended to remain within the Congress constellation, in contrast to the social movements which mobilized from outside. In both the labour movement post-Marikana and the student movement, new formations emerged beyond the Congress constellation, though large rumps remained within.

ANC traditions are reproduced, refashioned, and occasionally ruptured. Thus, both the social movements and the community protests reproduced earlier repertoires strongly associated with Congress traditions of collective action and illegality. The social movements became explicitly critical of the ANC and adopted a new discourse focused on anti-neo-liberalization and, in some cases, socialism. The new union formations and the EFF, which emerged or gained momentum from the Marikana strikes, tended to present themselves as the real custodians of the Congress vision and symbols such as the Freedom Charter, but refashioning these to emphasize elements that had been abandoned by the ANC, such as nationalization and socialism.

On the other hand, the mine workers who led the Marikana strikes were in some ways the most transgressive of all, profoundly rupturing the most institutionalized regions in the movement landscape in their rejection of established procedures and institutions of collective-bargaining, imploding the Congress-aligned mining union and joining a union aligned with a Black Consciousness federation. While the #FeesMustFall student movement tended to reproduce some of the repertoires of mass struggle, it gave birth to the most distinctive new symbols and discourses of the four cases presented here, appropriating previously marginalized Black Consciousness traditions and transforming them with a nascent Black Feminism and a commitment to intersectionality, thus marking a sharp break with ANC traditions.

It is notable that connections or expressions of solidarity were seldom established between the different organizations or movements involved in these four mobilizations, with each remaining more or less focused on grievances and authorities characteristic of their own specific regions of the landscape. This is not to say they did not influence each other. The imagery of Marikana in particular established a life of its own, being appropriated by a range of movements, from the students to community protesters to the EFF. It is also notable that violent tactics by protesters and police tended to emerge at certain points in all of these struggles. The lack of concrete links does not prevent mobilizations from sharing repertoires, symbols, and tactics.

Turning to organizational forms, the rejection of formal organizational structures in favour of temporary, ephemeral, and 'horizontal' organization emerges most strongly across the community protests, Marikana and some currents within #FeesMustFall – though echoed also within the social movement debates over organizational form – which appears to be at least in part a response to the conviction that formal organizational structures are inevitably co-opted and institutionalized, or appropriated by elites. Fluid forms of organization are preferred, marking a further break with Congress traditions of durable, centralized, and national organizations. While these tend to be absorbed into or replaced by formal structures such as trade unions or student organizations, they leave behind them possibilities, practices, and repertoires, always available for appropriation at some future time.

The analysis presented here, while tentative and selective, suggests that the manifold forms of collective mobilization and contention that have emerged in the post-apartheid movement landscape demonstrate a complex mix of trends, some tending to conserve and reproduce the existing landscape shaped and laid down by the histories of the Congress constellation of organizations, others drawing on these but reconfiguring and innovating at the level of both symbols and discourses, and organizational form. The weight and attraction of the Congress organizations, symbols, and repertoires remain powerful, yet there is considerable innovation on its margins. Some of these produce new practices, some continue to simmer or leave dormant forms that may re-emerge in new struggles, while others die out or are erased.

Levels of transgression are signalled by responses from elites – and here Marikana attracted the most concentrated physical and symbolic onslaught from within the state and the Congress constellation. Responses to the student movement were mixed, combining symbolic delegitimation with concessions and attempts to reabsorb leadership. Indeed, attempts to delegitimate oppositional movements as 'third force' or 'counterrevolutionary' attempts to 'destroy' Congress or produce 'regime change' are the most consistent element in ANC responses, despite the fact that such movements are entirely legitimated by the Constitution. The movement landscape is thus deeply fractured in its origins, with the logic of the politics of liberation coming up against the logic of constitutional democracy, and while this produces contradictory strategies and responses by authorities, it also marks out tensions within the strategies of movements. The movement landscape thus appears as an unstable set of structures that is actively constituted, reconstituted, and contested, where some innovations establish new organizational nodes and repertoires or even become transgressive, while others are absorbed or erased.

One of the questions that emerges from this analysis is whether the Congress political tradition has been exhausted, or whether it still serves as a fecund source of ideas, symbols, and narratives that can nourish new struggles. It is as if the ANC colossus has begun to fracture and break apart, spewing fragments of itself across the landscape, each marked by its origins and history but also presenting new opportunities.

Notes

1. Karl von Holdt writes: an idea encountered in a conversation with Laurence Cox at the annual Manchester 'Alternative Futures and Popular Protest' conference in 2014; it is intriguing to see how similarly our thinking has evolved in parallel since that discussion, though with significant differences of emphasis – see the references above. 'Movement landscape' provides a richly productive metaphor, and I am deeply indebted to Cox for sparking this exploration.
2. From here on, we use the term 'Congress' to refer to the broader set of organisations and traditions at the centre of which is the ANC, and 'ANC' to refer to the organisation so named.
3. The term 'new social movements' as used here does not correspond with its prior use in the context of Europe and North America. In South Africa, it became popular after 1998 in reference to a number of movements that emerged after 1994 and outside the old political formations.
4. Our use of the term 'elite' in this context is a relative one. In poor communities, such elite layers include some who are distinguished by wealth derived from local businesses or insertion into important regional networks of patronage; others simply possess some political capital by virtue of their history as ANC activists and location in one or other of the ANC constellation of organisations, and thus have some potential to claim patronage or dispense it on a small scale.
5. The Congress movement of the 1980s developed a strategy of organising in multiple 'sites of struggle', including schools and universities, combining struggles for specific changes in each such site with a national struggle to end apartheid.
6. Government reduced subsidies to universities in the late 1990s. Institutions of higher learning began to operate increasingly along business principles, with some of the results including rapidly and continually increasing student fees, and the contracting out (or outsourcing) of functions considered to be 'support' rather than 'core' services to private companies (Naidoo, 2006, 2009; Pendlebury & van der Walt, 2006).

Acknowledgements

We would like to acknowledge the contribution made to our thinking by fellow researchers Malose Langa, Crispen Chinguno, and Ahmed Veriava, and the helpful comments of many colleagues, and three anonymous reviewers.

Disclosure statement

No potential conflict of interest was reported by the authors.

Funding

This work is drawn from research supported by the National Institute for the Humanities and the Social Sciences (South Africa), Ford Foundation, The Royal Norwegian Embassy, Pretoria, the CS Mott Foundation, and the National Research Foundation (South Africa).

References

Alexander, P. (2010). Rebellion of the poor: South Africa's service delivery protests – a preliminary analysis. *Review of African Political Economy, 37*(123), 25–40.

Alexander, P. (2012a). *Protests and police statistics: Some commentary.* Research Chair in Social Change, University of Johannesburg. Retrieved from https://www.researchgate.net/publication/305399004_Protests_and_Police_Statistics_Some_Commentary

Alexander, P. (2012b). Barricades, ballots and experimentation: Making sense of the 2011 local government election with a social movement lens. In M. C. Dawson & L. Sinwell (Eds.), *Contesting transformation: Popular resistance in twenty-first century South Africa* (pp. 63–80). London: Pluto Press.

Alexander, P. (2013). Marikana, turning point in South African history. *Review of African Political Economy, 40*(138), 605–619.

Alexander, P., & Pfaff, P. (2013). Social relationships to the means and ends of protest in South Africa's ongoing rebellion of the poor: The Balfour insurrections. *Social Movement Studies, 13*(2), 204–221.

Alexander, P., Sinwell, L., Lekgowa, T., Mmope, B., & Xezwi, B. (2013). *Marikana: A view from the mountain and a case to answer.* London: Bookmarks.

Ballard, R., Habib, A., & Valodia, I. (eds.). (2006). *Voices of protest: Social movements in post-apartheid South Africa.* Pietermaritzburg: University of Kwazulu Natal Press.

Barker, C. (2013). Class struggle and social movements. In C. Barker, L. Cox, J. Krinsky, & A. G. Nilsen (Eds.), *Marxism and social movements* (pp. 41–62). Leiden: Brill.

Barker, C., Cox, L., Krinsky, J., & Nilsen, A. G. (eds.) (2013). Marxism and social movements: An introduction. In C. Barker, L. Cox, J. Krinsky, & A. G. Nilsen (Eds.), *Marxism and social movements* (pp. 1–37). Leiden: Brill.

Bond, P. (2000). *Elite transition: From apartheid to neoliberalism in South Africa.* London: Pluto Press.

Burawoy, M. (2017). Social movements in the neoliberal age. In M. Paret, C. Runciman, & L. Sinwell (Eds.), *Southern resistance in critical perspective: The politics of protest in South Africa's contentious democracy* (pp. 21–35). London: Routledge.

Chinguno, C. (2013). *Marikana and the post-apartheid workplace order* (Working Paper 1). Johannesburg: SWOP.

Chinguno, C. (2015a). The unmaking and remaking of industrial relations: The case of Impala platinum and the 2012–2013 platinum strike wave. *Review of African Political Economy, 42*(146), 577–590.

Chinguno, C. (2015b). Strike violence in post-apartheid South Africa. *Labour, Capital and Society, 48*(1 & 2), 93–119.

Chinguno, C., Kgoroba, M., Mashibini, S., Masilela, B. N., Maubane, B., Moyo, N., … Ndlovu, H. (Eds.). (2017). *Rioting and writing: Diaries of Wits Fallists.* Johannesburg: SWOP.

Cox, L. (2016a). Studying movements in a movement-become-state: Research and practice in postcolonial Ireland. In O. Fillieule & G. Accornero (Eds.), *Social movement studies in Europe: The state of the art* (pp. 303–310). Oxford: Berghahn Press.

Cox, L. (2016b). The southern question and the Irish question: A social movement perspective. In Ó. G. Agustín & M. B. Jørgensen (Eds.), *Solidarity without borders: Gramscian perspectives on migration and civil society alliances* (pp. 113–131). London: Pluto Press.

Dawson, H. (2014). Patronage from below: Political unrest in an informal settlement in South Africa. *African Affairs*, 113(453), 518–539.

Dawson, H. (2017). Protests, party politics and patronage: A view from Zandspruit informal settlement, Johannesburg. In M. Paret, C. Runciman, & L. Sinwell (Eds.), *Southern resistance in critical perspective: The politics of protest in South Africa's contentious democracy* (pp. 118–134). London: Routledge.

Desai, A. (2002). *We are the poors: Community struggles in post-apartheid South Africa.* New York: Monthly Review Press.

Dlakavu, S. (2017). #Feesmustfall: Black women, building a movement and the refusal to be erased. In C. Chinguno, M. Kgoroba, S. Mashibini, B. N. Masilela, B. Maubane, N. Moyo, … H. Ndlovu (Eds.), *Rioting and writing: Diaries of Wits Fallists* (pp. 110–115). Johannesburg: SWOP.

Dwyer, P. (2004). *The contentious politics of the Concerned Citizens Forum (CCF).* Centre for Civil Society Research Report No. 27. Retrieved from http://citeseerx.ist.psu.edu/viewdoc/download?doi=10.1.1.515.4362&rep=rep1&type=pdf

Friedman, S., & Mottiar, S. (2004). *A moral to the tale: The treatment action campaign and the politics of HIV/AIDS. Globalisation, marginalisation and new social movements in post-apartheid South Africa project.* Durban: Centre for Civil Society and the School of Development Studies, University of KwaZulu-Natal.

Gabowitsch, M. (2017). *Protest in Putin's Russia.* Cambridge: Polity Press.

Hart, G. (2013). *Rethinking the South African crisis: Nationalism, populism, hegemony.* Scottsville: University of KwaZulu Natal Press.

Jacobs, C. A. (2017). The outcasts: No retreat, no surrender! In C. Chinguno, M. Kgoroba, S. Mashibini, B. N. Masilela, B. Maubane, N. Moyo, … H. Ndlovu (Eds.), *Rioting and writing: Diaries of Wits Fallists* (pp. 116–120). Johannesburg: SWOP.

Kgoroba, M. (2017). A closer look at the #EndOutsourcing protest at wits university: The other side of the #FeesMustFall protest. In C. Chinguno, M. Kgoroba, S. Mashibini, B. N. Masilela, B. Maubane, N. Moyo, … H. Ndlovu (Eds.), *Rioting and writing: Diaries of wits fallists* (pp. 126–129). Johannesburg: SWOP.

Langa, M. (ed.). (2017). *#Hashtag: An analysis of the #FeesMustFall movement at South African universities.* Johannesburg: CSVR.

Langa, M., & von Holdt, K. (2012). Insurgent citizenship, class formation and the dual nature of community protest: A case study of 'kungcatsha'. In M. Dawson & L. Sinwell (Eds.), *Contesting transformation: Popular resistance in twenty-first century South Africa* (pp. 80–100). London: Pluto Press.

Mabasa, A. N. (2017). Life of ideas in the #FeesMustFall 2015. In C. Chinguno, M. Kgoroba, S. Mashibini, B. N. Masilela, B. Maubane, N. Moyo, … H. Ndlovu (Eds.), *Rioting and writing: Diaries of Wits Fallists* (pp. 130–134). Johannesburg: SWOP.

Malabela, M. (2017). We are not violent but just demanding free decolonised education: University of the Witwatersrand. In M. Langa (Ed.), *#Hashtag: An analysis of the #FeesMustFall movement at South African universities* (pp. 132–147). Johannesburg: CSVR.

Marais, H. (2001). *South Africa – limits to change: The political economy of transformation* (2nd ed.). London: Zed Books.

Mashibini, S. (2017). University as a site of struggle: Contestation of ideas, space and leadership. In C. Chinguno, M. Kgoroba, S. Mashibini, B. N. Masilela, B. Maubane, N. Moyo, … H. Ndlovu (Eds.), *Rioting and writing: Diaries of Wits Fallists* (pp. 39–43). Johannesburg: SWOP.

McKinley, D., & Naidoo, P. (2004). New social movements in South Africa: A story in creation. *Development Update*, 5(2), 9–22.

McKinley, D. T. (2012). *Transition's child: The anti-privatisation forum (South Africa).* Johannesburg: South African History Archive. Retrieved from http://school.r2k.org.za/wp-content/uploads/2015/03/McKinley-2012.pdf

Mthombeni, A. (2017). A home away from home: The tales of 'Solomon Mahlangu House'. In C. Chinguno, M. Kgoroba, S. Mashibini, B. N. Masilela, B. Maubane, N. Moyo, … H. Ndlovu (Eds.), *Rioting and writing: Diaries of Wits Fallists* (pp. 44–51). Johannesburg: SWOP.

Mukwedeya, T. G., & Ndlovu, H. (2017). Party politics and community mobilization in Buffalo city, East London. In M. Paret, C. Runciman, & L. Sinwell (Eds.), *Southern resistance in critical perspective: The politics of protest in South Africa's contentious democracy* (pp. 107–117). London: Routledge.

Municipal IQ. (2012). *Municipal IQ's municipal hotspots results 2012: 2012 tally; A violent and diverse year for service delivery protests.* Press release.

Municipal IQ. (2017, February 1). *2016 Figure: Service delivery protests suggest election year lull.* Press release. Retrieved from https://www.municipaliq.co.za/publications/press/201702010920416649.doc

Naidoo, P. (2002). From WCAR to WSSD: The united nations, globalisation and neoliberalism. *Khanya: A Journal for Activists, 1.*

Naidoo, P. (2006). 'Constituting the class': Neoliberalism and the student movement in South Africa. In R. Pithouse (Ed.), *Asinamali: University struggles in post-apartheid South Africa* (pp. 51–68). Trenton, NJ: Africa World Press.

Naidoo, P. (2009). Taming the young lions: The intellectual role of youth and student movements after 1994. In W. Gumede & L. Dikeni (Eds.), *The poverty of ideas: South African democracy and the retreat of intellectuals* (pp. 153–168). Johannesburg: Jacana.

Naidoo, P., & Veriava, A. (2005). Re-membering movements: Trade unions and new social movements in neoliberal South Africa. In *Centre for civil society, from local processes to global forces* (pp. 27–62). Durban: University of KwaZulu-Natal.

Naidoo, P., & Veriava, A. (2013). Predicaments of post-apartheid social movement politics: The anti-privatisation forum in Johannesburg. In D. Pillay, J. Daniel, P. Naidoo, & R. Southall (Eds.), *New South African review 3: The second phase – tragedy or farce?* (pp. 76–89). Johannesburg: Wits University Press.

Ndlovu, H. (2017). The journey through wits #FeesMustFall 2015/16. In C. Chinguno, M. Kgoroba, S. Mashibini, B. N. Masilela, B. Maubane, N. Moyo, … H. Ndlovu (Eds.), *Rioting and writing: Diaries of Wits Fallists* (pp. 30–38). Johannesburg: SWOP.

Pendlebury, J., & van der Walt, L. (2006). Neoliberalism, bureaucracy, and resistance at wits university. In R. Pithouse (Ed.), *Asinamali: University struggles in post-apartheid South Africa* (pp. 79–92). Trenton, NJ: Africa World Press.

Roseberry, W. (1994). Hegemony and the languages of contention. In G. M. Joseph & D. Nugent (Eds.), *Everyday forms of state formation: Revolution and the negotiation of rule in Mexico* (pp. 355–366). Durham: Duke University Press.

Runciman, C . (2012). Resisting privatisation: exploring contradictory consciousness and activism in the Anti-Privatisation Forum. In M. Dawson & L. Sinwell (Eds.), *Contesting transformation: Popular resistance in twenty-first century South Africa* (pp. 80–100). London: Pluto Press.

Runciman, C. (2015). The decline of the Anti-Privatisation Forum in the midst of South Africa's 'rebellion of the poor'. *Current sociology, 63*(7), 961–979.

Runciman, C. (2017). South African social movements in the neoliberal age. In M. Paret, C. Runciman, & L. Sinwell (Eds.), *Southern resistance in critical perspective: The politics of protest in South Africa's contentious democracy* (pp. 36–52). London: Routledge.

Runciman, C., Alexander, P., Rampedi, M., Moloto, B., Maruping, B., Khumalo, E. & Sibanda, S. (2016). *Counting police-recorded protests: Based on South African police service data* (South African Research Chair in Social Change Report #2). Johannesburg: South African Research Chair in Social Change, University of Johannesburg.

Sinwell, L. (2013). The Marikana strike: The origins of a living wage demand and changing forms of worker struggles in Lonmin platinum mine, South Africa. *Labour, Capital and Society, 46*(1&2), 93–115.

Sinwell, L. (2015). 'AMCU by day, workers' committee by night': Insurgent trade unionism at Anglo Platinum (Amplats) mine, 2012–2014. *Review of African Political Economy, 42*(146), 591–605.

Sinwell, L., & Mbatha, S. (2016). *The spirit of Marikana: The rise of insurgent trade unionism in South Africa.* London: Pluto Press.

Van Heusden, P., & Pointer, R. (2005). Subjectivity, politics and neoliberalism in post-apartheid Cape Town. In A. Alexander & M. Mbali (Eds.), *Problematising resistance* (pp. 125–166). Durban: Centre for Civil Society.

Veriava, A. (2015). Introduction: Reopening the constituent process. *South Atlantic Quarterly, 114*(2), 426–435.

Von Holdt, K. (2003). *Transition from below: Forging trade unionism and workplace change in South Africa.* Pietermaritzburg: University of Natal Press.

Von Holdt, K. (2014). On violent democracy. *The Sociological Review, 62*(2), 129–151.

Von Holdt, K., Langa, M., Molapo, S., Mogapi, N., Ngubeni, K., Dlamini, J., & Kirsten, A. (2011). *The smoke that calls: Insurgent citizenship, collective violence and the struggle for a place in the new South Africa; eight case studies of community protests and xenophobic violence* (Research Report). Johannesburg: CSVR and SWOP.

Zibechi, R. (2012). *Territories in resistance: A cartography of Latin American social movements.* Oakland: AK Press.

Uncovering a politics of livelihoods: analysing displacement and contention in contemporary India

Gayatri A. Menon and Aparna Sundar

ABSTRACT

In this paper, we use livelihoods as an organizing concept which brings together questions of production, social reproduction, and the conditions for these, to describe and reflect upon three 'moments' of displacement and contention in India. Our first moment, a massive flash strike by workers in the export garments industry in Bangalore, is located in the present neo-liberal context of jobless growth, increasingly unregulated and precarious forms of employment, and market-based forms of service provision. Our second moment concerns popular struggles in defence of the commons in settled rural fishing communities in south India, and the third, the tenacious efforts of pavement dwellers in Bombay to make place, the political condition for production and social reproduction. The originary context for these last two moments was the state-led, technology-driven, capitalist modernization of agriculture and fisheries of the early post-'independence' decades, tied to projects of state-building, self-reliance, and sovereignty. The three moments chart the long history of processes of precarization under postcolonial capitalism but, equally, a constant politics of livelihood, grounded in claims to rights earned through labour, and addressing itself to both state and capital as complicit in structuring access to livelihoods under capitalism.

Introduction

Two themes run through the growing literature on precariousness. The first is its origins in processes of neo-liberalization: accumulation by dispossession (Harvey, 2003), failed agrarian transition and agrarian distress (Bernstein, 2009; Vakulabharanam & Motiram, 2011), informalization of work, and jobless growth (Kannan & Raveendran, 2009), all of which have rendered ever-larger populations surplus to capital (Li, 2010; Sanyal, 2007). With dispossession from the direct means of subsistence, the failure of the generalized capitalist system to provide a living wage, and the retreat of state provision even in the heartlands of the capitalist welfare state, precariousness has come to be seen as the defining condition of the present (Standing, 2014, with its emblematic figures of the pauper (Breman, 2016), the migrant 'wage hunter and gatherer' (Breman, 1994), and the vagabond (Katz, 2001). The second theme has to do with the 'politics of precarity' (Lee & Kofman, 2012): the urgency of the 'social question' (Breman & van der Linden, 2014) and of projects of 'making live' (Harris & Scully, 2015; Ferguson, 2015; Li, 2010), and the difficulty of imagining what the social forces might be for such a project given the declining ability of organized waged labour to play this role (Scully, 2016). Tania Li posed this political question most acutely when she wrote: 'If the population rendered redundant to capital's requirements is to live decently, it will be because of the activation of a biopolitics that places the intrinsic value of life – rather than the value of people as workers or consumers – at its core. But what are the forces that would activate such a politics? And why would they do so?' (Li, 2010, pp. 67–68).

In seeking to locate such a politics, the first step is to remind ourselves that while the scale of lives rendered precarious is perhaps unprecedented, precarity itself is not a new development. Instead, it is a long-standing dimension of capitalism, co-originating with capitalism as a mode of production (Denning, 2010; Feldman, Menon, & Geisler, 2011) and produced for smaller or greater numbers of people by the different accumulation cycles and strategies of capital (Bernstein, 2007, p. 2). Indeed, it was only the interregnum of the welfare state that allowed us to forget its inevitability within capitalism (Neilson & Rossiter, 2008). Attention to the specific ways in which precarity is produced and resisted over this longer history would require that we examine what resources people draw on, or hold on to, in order to resist being made precarious; for what is ultimately at stake is not wages, or employment, but the means of living. We thus turn from a focus on precarity to what might be seen as its obverse – livelihoods.

We use livelihoods as an organizing concept that brings together questions of production, social reproduction, and the conditions for these, to chart three 'moments' of displacement and contention. These moments are drawn from our own past and present research; of them, only the first is located in the present neo-liberal context of jobless growth, increasingly unregulated and precarious forms of employment, and market-based forms of service provision. This is a recent massive flash strike by women workers in the export garments industry in Bangalore, India's third largest and premier IT city, to protest changes to the state-provided Provident Fund. We then work backward to trace a genealogy of this moment through two others. The second moment is that of popular struggles in defence of the commons in settled rural fishing communities in south India. In the third moment, we turn our attention to pavement dwellers in Bombay who trace their arrival in the city to displacement from agrarian livelihoods and place, and describe their tenacious efforts to make place, the political condition for production and social reproduction. The originary context for these last two moments was the state-led, technology-driven, capitalist modernization of agriculture and fisheries of the early post- 'independence' decades, tied to projects of state-building, self-reliance, and sovereignty.

In considering the current conjuncture and its shaping of the space and subject of politics, the struggles described here offer four kinds of insights. First, as moments of resistance, tied to imminent loss of livelihood, they articulate the meaning and scope of particular livelihoods. We see that it is not only work, or the income from it, that is being defended, it is also the conditions that make both work and income possible. Evoking Arendt's (1951) description of citizenship as 'the right to have rights', the struggles illustrate the conditions necessary to make a life. Second, if, for Arendt, the necessary condition of political rights is membership in a political community, the claims articulated by these struggles, although often addressed to the state, are not dependent on it; the conditions of livelihoods they name are firmly material, and produced by particular histories of labouring, in particular places; they are material rights earned through labour, rather than abstract rights derived from political inclusion. Third, in contrast to arguments that, by addressing themselves to the state rather than to capital, such struggles potentially relieve particular capitals of their responsibility for the loss of livelihoods, we see them instead as interrogating the compact between state and capital that, in the name of 'development', shapes access to livelihoods under capitalism. Fourth, we see that the politics of livelihoods is a constant politics, able to activate and draw upon local wellsprings of anger and courage, and mobilize smaller or larger networks of solidarity. As the conditions of livelihoods vary, so do the means by which they are eroded or removed, and the resources and relationships available to resist their erosion. While we do not claim to answer Li's question fully, and to predict what kinds of forces might come together to render the growing precarity politically urgent, we contend that to specify and articulate (in both senses of the word) the shared claims underlying these apparently disparate struggles is itself a step toward such a rendering.

In the next section, we outline the debates around the possible politics of labour under neo-liberal capitalism against which we position our arguments, and propose a conceptualization of livelihoods as a way to move past these debates. This is followed by three empirical sections, each focused on a single moment of contention. In the final section we discuss what is gained by uncovering a politics of livelihood.

Locating livelihoods within a politics of labour

A recurring theme in the discussion on the politics of labour within neo-liberal capitalism in India is the conjunction of the diminishing power of an increasingly informalized workforce vis-à-vis capital, versus the continuing electoral and moral power of the poor as citizens in legitimating a liberal democracy constitutionally committed to socio-economic transformation. An influential theorization that weaves these themes together is that by the late economist Kalyan Sanyal. Sanyal (2007; see also the related argument by Chatterjee, 2008) notes that a peculiar feature of postcolonial neo-liberal capitalism, not shared by the earlier capitalisms of the North, is the vast numbers of those made redundant to capital. The low demand for labour due to technological improvements and techniques of intensifying labour extraction, such as contracting out, means that those displaced from the land by processes of accumulation by dispossession are no longer absorbed into the economy as wage labour, or even as a reserve army of labour. As a result, contemporary postcolonial capitalism is composed of two spheres – the sphere of capital, organized around the logic of accumulation, and the sphere of non-capital, where the dominant logic is that of 'livelihood needs'. Since the mass of people in the sphere of 'non-capital' are redundant to capital and no longer relevant to its strategies, capital is indifferent to their ability to reproduce themselves. They are therefore unable to win concessions from it through the usual modes of labour bargaining. Modern-day conditions of political legitimacy, however, mean that the state is required to engage in a politics of 'making live' through minimal welfarist programmes, rendering those in the space of 'non-capital' as 'populations' and objects of governance, rather than political subjects able to make claims as citizens.

Critics have pointed to the difficulty of making this accumulation versus need, or capital versus non-capital, distinction, pointing out that many workers in the informal sector who seem irrelevant to capital are in fact employed in 'hidden wage relationships' through extended value chains (Roy-Chowdhury, 2014), and that most workers, in fact, 'circulate though different systems of work, as well as between work and non-work' (Gidwani & Wainwright, 2014, p. 45). But while few contemporary scholars agree with Sanyal's (and Chatterjee's) capital–non-capital, or population–citizen binaries, they tend to retain a distinction between the politics of (formal) labour that addresses itself to capital with demands around wages and working conditions, and the citizenship politics of informal labour that addresses itself to the state with demands around welfare, housing, services, and so on. Describing the politics of *beedi* (leaf-rolled cigarettes) and construction workers, who, given the extended value chains through which they are employed and the absence of a shared workplace, are unable to organize collectively against a common employer, Agarwala (2013) suggests that they instead address their demands for better working conditions to the state through a politics of citizenship. This politics of citizenship is also a feature of poor people's mobilizations resisting slum clearance and dispossession, as Bhan (2014) shows (although, as he also notes, the moral claim of poverty on citizenship is waning). While both Agarwala and Bhan (albeit the latter far more guardedly) see the resort to the language of citizenship as creating openings for informal workers and the urban poor more generally, RoyChowdhury (2014) argues that the language of citizenship deployed by the urban poor to demand land title in slums or other such basic needs works to disguise the structural nature of their poverty and deprivation and its roots in the exploitation of their labour. Given that many of them are working for multinational capital through extended value chains, it lets capital get away with not paying a living wage. Vijayabaskar (2011) makes a related argument in his work on labour in Tamil Nadu, where the state's successive populist governments have increased general welfare provision to its citizens while simultaneously seeking to retain capital by making significant concessions around wages and working conditions.

The use of livelihoods as an analytical category provides a way to reconcile these arguments about the politics of labour versus the politics of citizenship. We draw on two sources for our conceptualization. The first is the livelihoods framework that emerged in the 1990s within development research and practice (Scoones, 2009, p. 174). This defines 'a livelihood system' as comprising 'the capabilities, assets (including both material and social resources) and activities required for a means of living' (Carney, quoted in De Haan, 2012, p. 347). Livelihoods rarely refer to a single activity but include 'complex, contextual, diverse and dynamic strategies developed by households to meet their needs' (Gaillard, Maceda, Stasiak, le Berre, & Espald, quoted in De Haan, 2012, p. 347). The framework's focus on the household as the unit of study is helpful in making visible how members pool incomes from diverse sources, such as farming, wage work, and petty vending. The household is also the link to the second source for our conceptualization of livelihoods: feminist social reproduction theory. Social reproduction refers to 'the activities … responsibilities and relationships directly involved in the maintenance of life on a daily basis, and intergenerationally' (Laslet and Brenner, quoted in Luxton, 2006, pp. 35–36; see also Federici, 2008), such as the provision of food, clothing and shelter, the care and socialization of children, the care of the elderly and infirm, and the social organization of sexuality. The elements and processes that constitute social reproduction are not static, but historically, culturally, and socially produced; it refers to a 'standard of living' and not simply 'a bundle of commodities' (Picchio, 2000, p. 213). Bringing together livelihoods and feminist social reproduction theories gives us a conception of livelihoods that combines production and social reproduction activities and strategies, including diversification and income pooling, as well as the conditions, resources, and relationships that make these activities possible. The genealogy we trace in our empirical sections makes this conception of livelihoods both more concrete and more capacious. It includes a way of life; a means of living, a source of maintenance and sustenance; a wage or income; property (with security of ownership or tenure) in the form of space in which to live and work, a home, land as homestead or as source of small rents; other kinds of property necessary to self-employment or petty commodity production, such as tools, a vehicle; access to common property guaranteed by custom or law; state provision of income and basic needs; and forms of social support such as family and kin relationships.

As an analytical construct, livelihoods go some way in opening up the study of labour under neoliberal capitalism. First, it allows us to 'decentre wage labour in our conception of life under capitalism' (Denning, 2010, p. 80; see also Scully, 2012). Wage labour becomes central when other means of livelihood no longer exist, but it has never been the sole, or even the dominant, means of livelihood in the global South. Thinking in terms of livelihoods also provides an alternative to the distinction between the formal and informal sector that structures the debates outlined above. Given that some 93% of all Indian workers are located in the 'informal sector' (NCEUS, 2007), it might be wise to abandon the hopeful temporality of the term and to see it not as a sector in transition toward for-malization, but as a pervasive condition from which a lucky few escape, and not always permanently. As numerous empirical studies have shown (from Holmström (1984) to Cross (2010), and much of Jan Breman's oeuvre), the formal–informal distinction is a porous one. Members of a household may be spread across both sectors, or the same worker may be engaged in both kinds of work during the course of a working week, supplementing work in a regulated factory with paid domestic work, for instance (see also Scully, 2016). Social reproduction, likewise, relies on the interaction of various institutions, such as the state, the market, the family/household, and the third sector. While the social reproduction of 'classes of labour' (Bernstein, 2007, p. 2) involves social relations beyond direct capi-tal–labour relations, 'these relations are simultaneously shaped by, and exist in, articulation with the capitalist dynamic of accumulation' (Ferguson, 2016), so that shrinking state provision, or migration as a result of dispossession, can shift the burden of social reproduction work between different mem-bers of the household, or force households to turn to the market for needs, such as for food prep-aration or childcare. This reality of pooling resources, and of circulating between sectors, also undermines Sanyal's (2007) distinction between economies of accumulation and economies of need.

We now turn to an account of three spatially and temporally disparate moments of displacement and contention, tracing through them a genealogy of livelihoods and its politics.

Moment 1. Livelihoods and social security: garment workers' protests against Employee Provident Fund reforms, 2016[1]

This moment concerns a large-scale walk-out and demonstration mounted by garment workers in Bangalore against proposed government reforms to the terms of the Employee Provident Fund (EPF). We use it to explore the role of forms of state social security tied to waged work in social reproduction.

On 18 April 2016, some 120,000 workers, 80% of them women, walked out of the units of three major garment companies in Bangalore as well as several smaller units, to protest changes by the central government to the rules governing the EPF. Union leaders, who were taken by surprise, described the protests as 'spontaneous', a 'flash-strike', and 'leaderless'. The police commissioner who dealt with the protesters is reported as saying that when he asked a few leaders to come forward to negotiate, none did: 'When we asked them to submit a memorandum for us to take it forward, the women just said they wanted their PF money, that they wanted justice, he said' (Bageshree & Bharadwaj, 2016). According to a union leader from one of the major central trade unions, there had not been protests of this magnitude in any industrial sector in Bangalore in the last 10 years. The protests grew more militant over time, with tactics that included the blockade of a major arterial road. After three days, the government backed down and withdrew the proposed change.

The EPF is a savings instrument that comprises a monthly contribution by both the employer and the employee of 12% of the employee's basic salary. The current practice was that workers who were unemployed for two months could withdraw their entire provident fund savings. Under the new rules, to go into effect at the end of April 2016, workers would be able to withdraw only their con-tribution and the interest accrued on it; the employer's contribution could only be withdrawn at the retirement age of 58 (Yadav, 2016).

The Ready Made Garment (RMG) sector is a major employer in Bangalore, contributing as much as the IT sector, but far less well known in association with the city (RoyChowdhury, 2014). All the big brand names, such as Gap, H&M, Old Navy, Banana Republic, and JC Penny, source from here. There are estimated to be some 1200 big, small, and medium-size garment factories employing over

500,000 workers, 85% of them women (Yadav, 2016). Across Asia, the garments industry is heavily 'feminized', 'flexibilized', and amongst the most 'globalized' (Kumar, 2014, p. 790; Mezzadri, 2016). The RMG industry in India is facing competition, particularly from Bangladesh and Vietnam, and its industry body had sought an amendment in the laws to permit an increase in working hours in order to cater to peak season demand, and to 'compensate for lower labour productivity' (Sen, 2016). This was subsequently granted as part of a package of support to the industry announced later in 2016. Even prior to this announcement, the government had agreed to bear a major share of the employer's contribution to the EPF; it is likely the proposed changes were linked to the government's interest in delaying having to pay out the sums (Sen, 2016). Lankesh (2017) suggests that another reason was government's desire to limit frequent withdrawals in order to enable the vast central provident fund corpus to be used for speculative investment.[2]

Wages in the industry are low – averaging INR 8000 a month (Yadav, 2016) – and working conditions are hard, with high production targets and difficult relationships with the management. The factories run under the 'chain system' in which many tailors stitch different parts of the same garment. A group of 30–35 workers stitch a shirt every 40 seconds, and 70 workers stitch a pair of cargo pants every minute. The gruelling schedule leaves workers with little time for water or toilet breaks (Yadav, 2016). Overtime work, often involuntary, and penalties for not meeting production targets are very common. Sexual harassment is one of the biggest issues confronting women workers. Although two relatively new unions – the Garment and Textile Workers' Union (GATWU) and the Garment Labour Union (GLU) – have made some wage gains for the sector, activists note that it is very hard to organize the workers, as factory owners threaten to shut down factories if they see any sign of unionization, or harass workers who are known to be organizing. Even the largest of the unions, GATWU, has barely 8000 members across several units. The workers are predominantly young, a majority of them recent migrants from rural areas neighbouring the city and, increasingly, are brought in from parts of North and East India and housed in hostels near the factories. (The women who protested were not from this latter group, whose movements are tightly controlled, but had families in the city or its surrounding regions; men from their communities, many of them employed even more precariously than the women, as auto-rickshaw drivers or construction workers, were also visible in the protests; their demands were amplified by the region's Kannada-language media.) There is a high degree of labour turnover in the industry. A GATWU leader noted that while the majority of workers on the payrolls are permanent in order to comply with requirements of international apparel buyers, workers change jobs and factories also hire and fire workers frequently (Yadav, 2016). It is common for workers to be let go after four years to enable employers to avoid gratuity commitments and bonuses which kick in after a worker has completed five years at a unit. Mezzadri (2016, p. 1889) notes that this is one way of ensuring the disposability of women workers.

It is from this uncertain economic context, and in the shortened time horizons of workers with low expectations of health, vitality, or indeed, a long life, that the EPF derives its meaning and significance in the lives of garment workers. Far from a retirement fund, it is used by them to gain much-needed rest, tide over spells of unemployment as they move from job to job, and as a savings facility that they can draw on to pay for their children's education, illnesses and marriages (PUCL & WSS, 2017). An ILO survey found that nearly 80% of the sampled workers had accessed the EPF (cited in Sen, 2016). Some of the women quoted in the story claimed that they would not live until 58, so preventing them from accessing this amount was tantamount to denying them a share of their earned wages. A worker interviewed by one of our students said, 'you cannot work after 40 in such factories, your back will give up on you'; his life-history showed a break in employment every four or five years. A news story on the strike reported: 'The workers say that they work for their children's education, and that if they can't withdraw the full PF until they are 58, the money is of little use to them' (TNM Staff, 2016). Union activists we interviewed said:

> In a garment workers' family, if a child is studying in SSLC [Secondary School Leaving certificate, achieved after 10th grade] and gets good marks, the mother will quit the job because if she quits the job, she can get the money and she can join [enrol] her child in a good college [because] for admission in college, she will need to pay at least 30,000 to 40,000 rupees.

> They use the PF as a rotation and they don't want a long-term benefit.

Moment 2. Livelihoods and the commons: protests against mechanized trawling, 1995[3]

We describe here one instance of the struggle waged through the closing decades of the past century by fishers in the village of Kanyakumari (KK) in Tamil Nadu state to limit and regulate trawling. The villagers argued that the mechanized trawlers captured a disproportionate share of the resource, leading to overfishing and threatening the livelihood of the majority who fished on artisanal, largely non-mechanized craft. At stake was an understanding of the marine commons as a source of collective subsistence, and the village's customary right to govern access to these commons.

In early 1995, during an ongoing episode of contestation between the trawler owners and *kattumaram* (traditional craft) fishers of the village, the KK village committee put up a stone inscription outside the village church declaring a village law imposing strict regulation on trawling, to be observed by trawler operators in KK and five contiguous villages. In response to the trawler owners' violation of this law, over a thousand women from the village blocked the main highway running down to the cape with nets and *kattumarams* for three days, many with babies in their arms. The women did this, they said, because the men would have been arrested for similar action. After they ended their protest, a delegation of women went to the state capital of Chennai to call on the fisheries minister, to whom they claimed to have said: 'Our *kalvettu* [stone inscription] is the law in our village. We can't change it. If you change it, your law will remain in your office, it can't be implemented in our village'. They warned him what a decision against them could mean in terms of political support for the party, as the fishing villages of the district had always been an ADMK[4] *kōṭṭai* (fortress).When, a few days later, the trawlers did go out for a day, the *kattumaram* fishers issued a village writ threatening any merchant who bought fish from the trawlers with the denial of church rites. The owner of a large processing and marketing company, who belonged to the village, obeyed the writ because he had aging parents who might have needed their last rites. One trawler owner was forced to sign his agreement with the village when his mother died and was refused burial by the village committee. Despite there being no provision in church rules permitting such a denial, the parish priest was unable to act against his village committee's stand. Over the next few months, several more trawler owners agreed in writing to observe the regulations demanded by the village.

These events were responding to a fisheries development policy that was, in principle, intended to improve fisher livelihoods. Fishing, traditionally a caste-based occupation, had historically provided its practitioners only a subsistence-level livelihood. In the absence of freezing and transportation technology, markets for fresh catch were largely local; the sizable profits from the export of dried fish accrued largely to the traders. Merchants were also the beneficiaries of the seasonal and even daily vagaries of fishing, which kept fishers in constant need of loans to meet both consumption and production needs. Fisheries development planning began in the early 1950s with a vision of community development that included institutional reform in the form of cooperatives aimed at undermining the power of the merchants. Rather quickly, however, the imperatives of economic development and food production led to a shift, by the end of the first Five Year Plan (1951–1956), to the productivist, technocratic bias which underlay, for instance, the Green Revolution in agriculture (Kurien, 1992, p. 224). From the end of the First Five Year Plan, the primary means to modernization and increased productivity was the introduction of mechanized craft, for which the largest outlays in the fisheries sector were made. With the opening up of US and then Japanese markets for frozen lobster and prawn from the early 1950s, prawn exports increased exponentially, and the government began to provide subsidies for the acquisition of mechanized craft to supply these markets. While the growth in demand meant that all types of fishers were able to tap into it, the mechanized trawler owners were able to capture the largest share. Furthermore, despite being equipped, and required, to go out into deeper waters, they persisted in fishing inshore where the richest grounds were, indiscriminately scooping up all catch, including in the monsoon spawning season, and frequently destroying artisanal fishermen's nets in the process.

By the late 1970s, opposition to the trawlers had begun to flare in several states across India's coastline. This led to the emergence of a national coalition calling for the spatial and seasonal regulation of trawlers which saw success, in the form of a Marine Fisheries Regulation Bill and, by 1985, in the establishment of a national organization, the National Fishworkers' Forum (NFF). The harm caused to the resources and to the artisanal fishers by trawling and other types of indiscriminate mechanized fishing was for a long time the burning issue for the national fishworkers' movement. But it also from the outset challenged what it called 'the growth model of development', of which state support for mechanization was emblematic, which it argued resulted in displacement and dispossession (Dietrich & Nayak, 2006). Although the NFF had managed to establish a tenuous organizational presence in KK district, the actions described here drew little on its organizational strength, but rather on the authority of the customary village community, the use of church sanctions, and the villagers' assertion of their weight as voters. In part, this may be because, as sceptical observers of movements like the NFF have argued, 'there is a difference between the people's perception of what they are fighting for – basic subsistence denied by the state, and the claims made by intellectuals who postulate that indigenous resistance is a comprehensive critique of development' (Baviskar, 1995, p. 237; Nilsen, 2010).

To suggest that what was at stake was a 'mere' defence of subsistence, however, is to fail to see the larger assumptions and claims contained in commons-based subsistence. Frequently heard during the struggle described here and on several similar occasions across the district was the question: 'Don't we *all* need to live?' Villagers spoke proudly of their long history as skilled fishermen in the region, arguing that this gave them an intimate knowledge of the local ecology and its seasonal variations, and the labour processes and technologies best suited to them. The sovereignty of village law and its enforcement through church rites derived from, and upheld, this long association of space, residence, and labour (see Subramanian, 2009). This set of claims bears striking resemblance to the conception of the commons that Linebaugh (2008, pp. 44–45) interprets as embodied in the earliest version of the Magna Carta: commons rights were 'embedded in a particular ecology with its local husbandry', 'entered into by labor', and upheld by the sovereign laws of the collective that depends upon them for its livelihood.

Moment 3. Livelihoods and the right to life: Pavement Dwellers' case, 1985[5]

This third moment concerns the routine – but also the occasionally spectacular – evictions of the poor from the city and the demolitions of their dwellings, workplaces, and communities. We analyse the spatial politics and justice that govern the lives of the urban poor in India by focusing on the 1985 Supreme Court of India judgment on a case concerning the large-scale eviction of the urban poor that took place in July 1981 in Mumbai, a case now known as the Pavement Dwellers' case. The case challenging the eviction of people living on the pavements of the city pivoted on interpreting the Right to Life as a right that rested on recognizing citizens' right to a space to earn a livelihood.

In Mumbai, the commercial centre of India, over half of the city's residents live in informal settlements. They are a vital but invisibilized part of the city's labour force, the vast majority contributing to the city's booming informal economy but a significant number of whom are also employed in the formal sector, in the lower ranks of private and public sector enterprises and government agencies such as the port, police, and municipality. The informal settlements are located on either public or private land, and their residents live in precarious physical and political circumstances. Their lives and livelihoods are charged by the insecurity of tenure that governs their cramped, densely packed, and environmentally degraded spaces of social reproduction that often double up as spaces of production. There is a wide range of informal settlements, from those that have some official recognition and are therefore relatively protected from demolitions and evictions, to those that have none and live under a looming shadow of violence.

Poorest amongst these insecurely housed populations are the pavement dwellers who occupy the lowest tier of the urban labour market and who are compelled to live on the sidewalks. Their lives and livelihoods speak directly to the extreme challenges to securing place in the city, the place to produce and to reproduce themselves. Their poverty is compounded by their invisibility with respect to the provision of public services, as they have to pay not only for the few square metres that they squat on, but also for access to basic amenities such as water, toilets, and electricity. Within this precariously positioned existence, a demolition is a devastating event, and occurs frequently. For someone whose livelihoods depends on re-selling watch straps, or wielding a pick-axe, or mending shoes, or sorting and re-selling scrap paper and plastics, the loss of these tools and resources in the course of an eviction, along with the loss of home, is hard to recover from.

While there is evidence of pavement dwelling in Bombay dating back to the nineteenth century (see Orr, 1917), the 1970s saw a sharp increase in the numbers of people migrating to the city and compelled to live in slums and on the pavement in order to survive (Swaminathan, 2003). The growing and visible presence of the urban poor, and of pavement dwellers in particular, coincided with and stymied the city's attempts to project itself as the next 'Asian Tiger'.

In a bid to make Bombay the next Bangkok, on 13 July 1981, the chief minister of Maharashtra directed his government to clear the city's pavements of the people who lived on them by 21 July 1981. What ensued was a demolition operation that was unprecedented in its scale and its planning. While pavement dwellings in the city have always been subject to frequent demolitions, these drives had been sporadic and piecemeal. Never before had the pavement dwellings of the entire city been the target of a single demolition drive. One hundred thousand people were affected by this exercise of beautifying the city (Shah, 1986).

While the decisive actions of the state government to 'clean-up' the city were applauded by several sections of society, activists and sections of the press were outraged by the sheer brutality with which this eviction drive was executed. Unlike the contemporary period where the media is often an active supporter of 'clean Mumbai' efforts – 'cleanliness' being a convenient euphemism for ridding the city of the poor and minorities – in the early 1980s, still smarting from the censorship of the Emergency period in the 1970s, the media was less forgiving of the state's actions against the poor, who also had been a target of some of the era's worst excesses. There was some ambivalence amongst housing activists about whether or not to approach the court to arrest the demolitions, their concern being that while it was essential to bring the demolitions to a halt, it was unlikely that the court would rule in favour of the rights of the poor to squat on pavements; furthermore, that the ruling would jeopardize their nascent efforts to gain secure tenurial rights for the urban poor. Civil liberties activists and a few journalists however used their eye-witness accounts of the demolitions and 'deportations' to file petitions for a Public Interest Litigation[6] in the Bombay High Court, claiming that the state was violating the constitutional right to live and work anywhere in India, and were able to get an injunction that required the state to suspend the 'deportations'. Additionally, a stay on the demolitions was won on humanitarian grounds, the monsoon being perceived to be a particularly inappropriate time to conduct demolitions. Meanwhile another journalist, Olga Tellis, filed a similar petition in the Supreme Court of India.

The Pavement Dwellers' case, as these petitions collectively came to be known, took four years of litigation before judgment was delivered. While emerging from different starting points, the petitions eventually converged on the challenge to the constitutionally mandated Right to Life that the evictions posed. This case marks an important moment in the politics of livelihood that this paper delineates. The petitioners' argument was that the evictions, by denying the urban poor the wherewithal to pursue a livelihood, that is a place to reproduce themselves as labour, constituted a denial of the right to life. The petitioners were comprised of pavement dwellers, slum dwellers, civil liberties activists, and journalists. We learn from the petitions that the former two sets of petitioners earned their livelihood as factory workers, daily wage workers, cobblers, construction workers, hotel employees and, in fact, as employees of the Bombay municipality. The descriptions of their work and earnings demonstrated to the court that while their labour was in demand, their earnings were not sufficient to acquire legally acceptable accommodation in the city. Furthermore, these insecurely housed members of the city's labour force lacked an alternative, for they did not have a place to go to, or to return to, in the event of expulsion from Mumbai.

The argument made by the advocates of the petitioners was that, given the state's failure to ensure the availability of low-income housing in the city, by denying pavement dwellers the right to live on the pavement, their ability to earn a livelihood was denied. The denial of the right to earn a living constituted a threat to the right to life.[7] It was an ingeniously argued case, and indeed a study of the ruling handed down by Justice Chandrachud reveals that the court was forced to recognize the validity of the argument made by the petitioners and concede that the Right to Life included the right to a livelihood. This 1985 judgment has been hailed as a landmark in Indian jurisprudence because, by recognizing that the deprivation of a livelihood was tantamount to the deprivation of life, it expanded the scope of the existing constitutional right to life (Bhushan, 2004; Dhavan, 1989).

Discussion: a politics of livelihoods?

We begin with the obvious, but necessary, point illustrated by the three moments: far from being novel to the neo-liberal present, processes of precarization have a long history within capitalism. The most extreme form of precarization we describe is also the earliest – that of the Mumbai pavement dwellers who were twice displaced, first from agricultural land by processes of capitalist differentiation, and then for purposes of urban accumulation. Both kinds of dispossession took place within the era of state-led developmentalism, but their logics were not unique to it and continue to repeat themselves in the contemporary period. This is true too of the more insidious process of enclosure and displacement faced by the KK fishing communities, where technology served as the vector by which state-led modernization ushered in global markets. More unique to the neo-liberal present, perhaps, are the logics underlying the restructuring of the Bangalore garment workers' EPF entitlements: the state's desire to use the EPF funds for speculative investment, and to reduce employers' obligations as a way of keeping India's garment sector competitive in a fast-shifting global marketplace.

In each moment, the nature of the refusal to being made precarious took different forms, tied to the nature of livelihood, but as significantly, to the actors' relationships to place and community. The pavements dwellers, displaced from long-settled community, and facing displacement from place, were least able to act autonomously, and relied most heavily on interventions by the media and lawyers. In contrast, the KK villagers represented themselves through the customary village committee, whose laws over the commons they considered sovereign. They asserted their power as voters and traditional ADMK supporters, bent the local church to their will, and suborned the entire community, men, women, and children, into action. Although a national movement existed, and had even prior to these events succeeded in winning marine fisheries legislation nationally, there was little reference to, or reliance on, the leaders and structures of this movement in the moment described. The garment workers, similarly, drew their strength from others like them with roots and links in the city and surrounding districts; they were supported by their families in taking over the city's streets, and their demands were amplified by Bangalore's Kannada-language media. Although unions existed within the sector, and had been active around workplace issues, including those of sexual harassment which were of particular concern to the women, in this instance the unions were taken by surprise.

Our interest in the moments we describe, however, is not what they reveal about political organization, alliances, and mobilization. Nor are we making the argument that the gains they achieved were absolute and enduring. In fact, despite the significant expansion of the right to life by making livelihoods central to it by the judgment on the Pavement Dwellers' case, the outcome of the court's deliberations was to legitimate the state's right to evict pavement dwellers without providing them with a place to live. All that the judgment conceded was that the municipality could not evict pavement dwellers without giving them notice. Since 1985, there have been numerous evictions of the urban poor – and not just in Mumbai – some of which have been at scales much larger than the 1981 demolition drive in Mumbai. Likewise, in Kanyakumari two decades on, most of those who opposed the trawlers now own them. And while the Bangalore garment workers were successful in preventing the restructuring of the EPF, their poor wages and working conditions remain unchanged. These moments interest us because they reveal most clearly the claims underlying the resistance to being made precarious, claims which are articulated by the protagonists, *and acknowledged* by the state in its immediate response.

Since each struggle was primarily defensive, it might be assumed that what was being defended was some minimal entitlement, rather than a more expanded set of demands from the state or capital, for higher wages, benefits, or new forms of social security, for instance. However, insofar as these were sui-generis claims, not tied to strategic calculations about how much a particular capitalist, or capital as a whole, or the state, might be willing to offer in a bargaining situation, implicit in each was a historically specific and potentially politically generative sense of the necessities of a stable life. Most startling, because it has hitherto been taken for granted, is the question of space raised by the pavement dwellers when they argued that space to live in and work from was essential to the pursuit of livelihoods. If space was the condition of production and social reproduction for the pavement dwellers, for the fishworkers it was the right to the commons as a way of ensuring ongoing collective sustenance. For the garment workers, the EPF was a condition of social reproduction, both by enabling periodic rest and recovery from the harsh discipline of the factories for the workers themselves, and by enabling their children's education or wedding expenses. It is through the language of livelihoods that these disparate claims – for space to work from, control over the natural resource commons, and social security tailored to the ongoing needs of daily and generational social reproduction – become commensurate and mutually intelligible. To designate these as livelihood claims is not to argue that *individually* any one of them constitutes the conditions for a good, or even an adequate, livelihood. Instead, we see them as *together* contributing a genealogy of meanings that give livelihoods its capacious and generative potential as an alternative to precariousness.

Claims such as these, about the conditions of life and social reproduction, even when tied to the production process itself, are often seen as Arendtian 'citizenship' claims addressed to the state rather than to capital, especially in a context where extended value chains mask lines of ownership (Agarwala, 2013, on informal sector workers; Bhan, 2014; Desai & Sanyal, 2012, on urban slum dwellers). But the language of citizenship takes away from the sense conveyed in our struggles, of material entitlements *earned* through labour rather than abstract rights *granted* through political inclusion. Furthermore, our struggles question the assumption of sovereignty associated with citizenship claims. This is illustrated most vividly by the fishworkers, for whom the sovereignty of the state was conditional and strategically invoked compared to the more immediate sovereignty of the village community, tied to its control over the commons and its right to grant or withhold church rites. If, while addressing themselves to the state, these actors do not grant it sovereignty, while not addressing themselves directly to capital, it is nevertheless the ground against which they struggle. We read their protests as indicting both state and capital, and the political compact between in the name of 'development' that structures access to livelihoods.

Our final point is a methodological one. In studying collective action, there is a tension between seeing politics as always already there and taking it seriously, on the one hand, and wanting to see more organized and concerted action commensurate with the scale of the devastation, on the other hand. On the one hand are those like Bernstein (2007, pp. 8–9), calling for organized political action by the classes of labour and their organizations; and on the other are Harney and Moten (2013, p. 19), who compel us towards an understanding of politics that, in waiting for organized action to emerge, does not invisibilize the resilient struggle to carve out a life and a livelihood:

> In the trick of politics we are insufficient, scarce, waiting in pockets of resistance, in stairwells, in alleys, in vain. The false image and its critique threaten the common with democracy, which is only ever to come, so that one day, which is only never to come, we will be more than what we are. But we already are. We're already here, moving. We've been around. We're more than politics, more than settled, more than democratic.

It is in the stubborn refusals to capitulate that shape the conditions of everyday life, always already in a state of emergency, that we see the generative potential of a politics of livelihood.

Notes

1. This section is based on ongoing work by the authors and other researchers in the Livelihoods Initiative at Azim Premji University, Bangalore.
2. Lankesh (2017) notes that a similar attempt had been made by the national government in 2000; then, too, women garment workers in Bangalore had come out in protest, and the changes had not gone ahead.
3. This section is drawn from Sundar (2010).
4. Anna Dravida Munnetra Kazhagam, one of the two major political parties in Tamil Nadu state, and in power in that period.
5. This section is drawn from Menon's (2009) doctoral research.
6. The Public Interest Litigation emerged as a legal facility in the late 1970s as courts attempted to make themselves more accessible to the nation's most vulnerable and exploited subjects: they relaxed the rule

of standing so as to allow petitions for the court's intervention to be made on the basis of a third party report such as a newspaper article on a particular instance of injustice (Shukla 2006).

7. Olga Tellis and others v. Bombay Municipality Corporation and others, AIR 1986 SC 180; see also Vayyapuri Kuppusami and others v. State of Maharashtra and others, All India Reporter 1986 SC 180.

Acknowledgements

This paper has benefitted considerably from the comments of its anonymous reviewers, as well as from Alf Nilsen's thoughtful suggestions. We would also like to thank our colleague Rajesh Joseph whose help and insight have been invaluable for the section on garment workers.

Disclosure statement

No potential conflict of interest was reported by the authors.

References

Agarwala, R. (2013). *Informal labor, formal politics, and dignified discontent in India*. Cambridge: Cambridge University Press.
Arendt, H. (1951). *The origins of totalitarianism*. New York: Harcourt, Brace and Company.
Bageshree, S., & Bharadwaj, K. V. A. (2016, April 20). Anatomy of a leaderless protest. *The Hindu*.
Baviskar, A. (1995). *In the belly of the river: Tribal conflicts over development in the Narmada Valley*. New Delhi: Oxford University Press.
Bernstein, H. (2007, March 26–28). *Capital and labour from centre to margins*. Paper presented at the 'Living on the Margins' Conference, Stellenbosch.
Bernstein, H. (2009). Agrarian questions from transition to globalization. In A. Akhram-Lodhi & C. Kay (Eds.), *Peasants and globalization: Political economy, rural transformation and the agrarian question* (pp. 239–262). New York, NY: Routledge.
Bhan, G. (2014). The impoverishment of poverty: Reflections on urban citizenship and inequality in contemporary Delhi. *Environment and Urbanization*, 26(2), 547–560.
Bhushan, P. (2004, May 1). Supreme court and PIL: Changing perspectives under liberalisation. *Economic and Political Weekly*, 39(18), 1770–1774.
Breman, J. (1994). *Wage hunters and gatherers: Search for work in the urban and rural economy of South Gujarat*. Delhi: Oxford University Press.
Breman, J. (2016). *On pauperism in present and past*. New Delhi: Oxford University Press.
Breman, J., & van der Linden, M. (2014). Informalizing the economy: The return of the social question at a global level. *Development and Change*, 45(5), 920–940.
Chatterjee, P. (2008, April 19). Democracy and economic transformation in India. *Economic and Political Weekly*, 43(16), 53–62.

Cross, J. (2010). Neoliberalism as unexceptional: Economic zones and the everyday precariousness of working life in South India. *Critique of Anthropology, 30*(4), 355–373.

De Haan, L. J. (2012). The livelihood approach: A critical exploration. *Erdkunde, 66*(4), 345–357.

Denning, M. (2010). Wageless life. *New Left Review, 66,* 79–97.

Desai, R., & Sanyal, R. (2012). *Urbanizing citizenship: Contested spaces in Indian cities.* Thousand Oaks, CA: Sage.

Dhavan, R. (1989, June 2). Putting the clock back. *Indian Post.*

Dietrich, G., & Nayak, N. (2006). Exploring the possibilities of counter-hegemonic globalization of the Fishworkers' movement in India and its global interactions. In B. de Sousa Santos (Ed.), *Another production is possible: Beyond the capitalist canon* (pp. 381–414). London: Verso.

Federici, S. (2008). *Precarious labor: A feminist viewpoint.* In the Middle of a Whirlwind: 2008 Convention Protests, Movement and Movements. Retrieved from https://inthemiddleofthewhirlwind.wordpress.com/precarious-labor-a-feminist-viewpoint/

Feldman, S., Menon, G. A., & Geisler, C. (2011). Introduction: A new politics of containment. In S. Feldman, C. Geisler, & G. A. Menon, (Eds.), *Accumulating insecurity: Violence and dispossession in the making of everyday life* (pp. 1–23). Athens: University of Georgia Press.

Ferguson, J. (2015). *Give a man a fish: Reflections on the new politics of distribution.* Durham, NC: Duke University Press.

Ferguson, S. (2016). Intersectionality and social-reproduction feminisms: Toward an integrative ontology. *Historical Materialism, 24*(2), 38–60.

Gidwani, V., & Wainwright, J. (2014, August 23). On capital, not-capital, and development: After Kalyan Sanyal. *Economic and Political Weekly, 49*(34), 40–47.

Harney, S., & Moten, F. (2013). *The undercommons: Fugitive planning and black study.* Wivenhoe: Minor Compositions.

Harris, K., & Scully, B. (2015). A hidden counter-movement? Precarity, politics, and social protection before and beyond the neoliberal era. *Theory and Society, 44*(5), 415–444.

Harvey, D. (2003). *The new imperialism.* New York, NY: Oxford University Press.

Holmström, M. (1984). *Industry and inequality: The social anthropology of Indian labour.* Cambridge: Cambridge University Press.

Kannan, K. P., & Raveendran, G. (2009). Growth sans employment: A quarter century of jobless growth in India's organised manufacturing. *Economic and Political Weekly, 44*(10), 80–91.

Katz, C. (2001). Vagabond capitalism and the necessity of social reproduction. *Antipode, 33*(4), 709–728.

Kumar, A. (2014). Interwoven threads: Building a labour countermovement in Bangalore's export-oriented garment industry. *City, 18*(6), 789–807.

Kurien, J. (1992). Ruining the commons and responses of the commoners: Coastal overfishing and fishworkers' actions in Kerala State, India. In D. Ghai & J. M. Vivian (Eds.), *Grassroots environmental action: People's participation in sustainable development* (pp. 221–258). London: Routledge.

Lankesh, G. (2017). The mirage of an oasis in old age. In C. Gowda (Ed.), *The way I see it: A Gauri Lankesh reader* (pp. 223–226). New Delhi: Navayana.

Lee, C. K., & Kofman, Y. (2012). The politics of precarity: Views beyond the United States. *Work and Occupations, 39*(4), 388–408.

Li, T. M. (2010). To make live or let die? Rural dispossession and the protection of surplus populations. *Antipode, 41*(s1), 66–93.

Linebaugh, P. (2008). *The magna carta manifesto: Liberties and commons for all.* Berkeley: University of California Press.

Luxton, M. (2006). Feminist political economy in Canada and the politics of social reproduction. In K. Bezanson (Ed.), *Social reproduction: Feminist political economy challenges neo-liberalism* (pp. 11–44). Montreal: McGill-Queen's University Press.

Menon, G. A. (2009). *Living conditions: Citizens, 'squatters', and the politics of accommodation in Mumbai* (Unpublished doctoral dissertation). Ithaca, NY: Cornell University.

Mezzadri, A. (2016). Class, gender and the sweatshop: On the nexus between labour commodification and exploitation. *Third World Quarterly, 37*(10), 1877–1900.

National Commission for Enterprises in the Unorganised Sector (NCEUS). (2007). *Report on conditions of work and promotion of livelihoods in the unorganised sector.* Retrieved from http://dcmsme.gov.in/Condition_of_workers_sep_2007.pdf

Neilson, B., & Rossiter, N. (2008). Precarity as a political concept, or, Fordism as exception. *Theory, Culture & Society, 25*(7–8), 51–72.

Nilsen, A. G. (2010). *Dispossession and resistance in India: The river and the rage.* New York, NY: Routledge.

Orr, J. P. (1917). *Social reform and slum reform.* 2 parts. Bombay: Times Press.

People's Union for Civil Liberties (PUCL) and Women Against Sexual Violence and State Repression (WSS). (2017). *Thread and tension: An account of the historic uprising of garment workers.* Bengaluru: PUCL and WSS.

Picchio, A. (2000). Wages as a reflection of socially embedded production and reproduction processes. In L. Clarke, P. de Gijsel, & J. Janssen (Eds.), *The dynamics of wage relations in the New Europe* (pp. 195–213). Dordrecht: Springer.

RoyChowdhury, S. (2014). Bringing class back in: Informality in Bangalore. *Socialist Register, 51*(51), 73–92.

Sanyal, K. (2007). *Rethinking capitalist development: Primitive accumulation, governmentality and post-colonial capitalism.* London: Routledge.

Scoones, I. (2009). Livelihoods perspectives and rural development. *The Journal of Peasant Studies, 36*(1), 171–196.

Scully, B. (2012). Land, livelihoods, and the decline of work: South African lessons for current debates. *Journal of World-Systems Research, 18*(1), 90–102.

Scully, B. (2016). From the shop floor to the kitchen table: The shifting centre of precarious workers' politics in South Africa. *Review of African Political Economy, 43*(148), 295–311.

Sen, C. (2016, May 13). Attention please: Policy lessons from the Bangalore garment workers' unrest. *The Telegraph.* Retrieved from https://www.telegraphindia.com/1160513/jsp/opinion/story_85222.jsp

Shah, A. (1986). Street life. *Imprint, 25,* 34–41.

Shukla, R. (2006). Rights of the poor: An overview of Supreme Court. *Economic and Political Weekly, 41*(35), 3755–3799.

Standing, G. (2014). Understanding the precariat through labour and work. *Development and Change, 45*(5), 963–980.

Subramanian, A. (2009). *Shorelines: Space and rights in South India.* Stanford, CA: Stanford University Press.

Sundar, A. (2010). *Capitalist transformation and the evolution of civil society in a South Indian fishery.* (Unpublished doctoral dissertation). University of Toronto, Toronto.

Swaminathan, M. (2003). Aspects of poverty and living standards. In S. Patel & J. Masselos (Eds.), *Bombay and Mumbai: The city in transition* (pp. 81–109). New Delhi: Oxford University Press.

TNM. (2016, April 18). Monday blues: Why thousands of garment factory workers blocked a major Bengaluru road. *The News Minute.* https://www.thenewsminute.com/article/monday-blues-why-thousands-garment-factory-workers-blocked-major-bengaluru-road-41800

Vakulabharanam, V., & Motiram, S. (2011). Political economy of agrarian distress in India since the 1990s. In S. Ruparelia, S. Reddy, J. Harriss, & S. Corbridge (Eds.), *Understanding India's new political economy: A great transformation?* (pp. 101–126). Hoboken, NJ: Taylor & Francis.

Vijayabaskar, M. (2011). Global crises, welfare provision and coping strategies of labour in Tiruppur. *Economic and Political Weekly, 46*(22), 38–45.

Yadav, A. (2016, April 22). Bengaluru protests represent a new wave of militant worker expression, say union leaders. *Scroll.in.* Retrieved from http://scroll.in/article/806968/bengaluru-protests-represent-a-new-wave-of-militant-worker-expression-say-union-leaders

A precarious hegemony: neo-liberalism, social struggles, and the end of Lulismo in Brazil

Ruy Braga ⓘ and Sean Purdy ⓘ

ABSTRACT

Analysis of Brazil's political and economic crisis tends to emphasize the economic 'errors' that President Dilma Rousseff's Workers' Party (PT) government inherited from her predecessor Luíz Inácio Lula da Silva. It is clear, however, that political regulation is too narrow a focus to understand the current crisis. Such an explanation is unable to reveal the changes in class structure that took place during the Lula era as well as the effects of the international economic crisis. This article identifies the limits of the Brazilian development model and the main features of Lula's mode of regulation; analyses the conflicts produced by the neo-liberal regime of accumulation and the Lulista mode of regulation, emphasizing the role of precarious work in the current historical cycle of strikes and popular struggles in Brazil; and, finally, interprets the palace coup promoted by the social forces behind the impeachment of President Rousseff.

In general, analysis of Brazil's current political and economic crisis emphasizes the economic policy 'errors' that President Dilma Rousseff's Workers' Party (PT) government inherited from her predecessor, Luíz Inácio Lula da Silva. If it is true that certain political decisions of the federal government interfered with the dynamics of and conflict around Brazilian distributive efforts,[1] it is abundantly clear that an exclusive focus on political regulation is too narrow to illuminate the complexity of the current crisis. This focus is unable to reveal the changes in class structure that took place during the Lula era (2003–2011), not to mention the effects of the international economic crisis. Indeed, analyses in this mould fail to explain how the relationship between political regulation and economic accumulation not only failed to pacify class conflict, but instead radicalized it.

We advance an alternative interpretation of Brazil's current economic and political crisis and the 2016 coup against the Rousseff government. Rather than considering the criminal investigations of the government by the federal police or the activism of public prosecutors in combatting corruption as linchpins, we argue that the crisis and the coup resulted from the inadvertent convergence and consolidation of three historical developments, leading to the installation of an authoritarian, illegitimate federal government that followed a drastic neo-liberal programme of cutbacks and attacks on labour and social rights.

It is important to underline that we are *not* stating that the disastrous economic policies, especially of Rousseff's second government (2015–2016), or the adoption of austerity measures, such as cuts in public spending during the deepening economic recession, did not play a role in the 2016 coup.

Indeed, one of the most insightful analysts of the PT governments has convincingly argued that Rousseff's economic choices and the pressures that the financial markets exerted on her government demonstrably intensified the economic crisis (Bastos, 2017). What we are arguing is that the coup was the result of *both* structural tendencies and immediate political and economic decisions. In particular, we contend that the collapse of the Lulista hegemony was caused to a significant degree by the radicalization of the distributive struggles between the social classes.

André Singer's (2017a, 2017b) examination of the federal police's criminal investigation of the second Rousseff government, conducted under the code name of 'Car Wash Operation', and of the heightened political protagonism of the Public Prosecutor's office has helped us understand the current moment as one marked by a crisis between the different powers in the republic. The judiciary stepped in to fill what it saw as a 'power vacuum'. However, this vacuum had been created by a combination of economic and political crises that were produced at the end of the first Rousseff government by the exhaustion of the Brazilian development model and the end of the consensus between the subaltern and the dominant classes.

As such, we first highlight the emergence of the discontent of workers, especially precarious workers, with the limits of the development model managed by the PT. The current crisis cannot be understood without taking into account the restlessness of organized workers who engaged in a sharp increase in strike activity in 2013 and 2014. Second, we define and develop the notion of the 'precarious hegemeony' of *Lulismo*, that is, the construction of both the 'passive' consent of the subaltern classes and the 'active' consent of the trade union bureaucracy. Third, we outline in detail the 'June Days' of 2013 when the young, urban precariat took to the streets against fare hikes in public transport and poor public services, definitively undermining the precarious hegemony of Lulismo and the PT governments' development model. Finally, we argue that the current crisis is the product of a notable adjustment in the functioning of the Brazilian development model led by the business classes through their imposition of drastic austerity measures aimed at cheapening the costs of the national labour force and deepening the current regime of accumulation by pillage. Each in their own way, these structural shifts contributed to the crisis of Lulismo and the subsequent consolidation of the current state of exception in the country.

Strike cycles

According to the latest data from the Strike Tracking System of the Inter-Union Department of Statistics and Socioeconomic Studies (SAG-DIEESE), Brazilian workers staged a historically unprecedented strike wave in 2013, totalling 2050 strikes (DIEESE, 2015). This meant an increase of 134% over the previous year and was the highest recorded to date. This high level in 2013 stands in contrast to low levels throughout the 2000s that had been shaped by what Marcel Badaró Mattos (2015a) calls the 'progressive pacification' of much of the combative union leadership and its incorporation into the PT governments (see also Galvão, 2014). In the 2010s, this marked decline was overcome and the trade union movement regained at least part of its political vigour. While detailed data on the strikes that took place in 2014 are not yet available, wage gains again surpassed those of the preceding year, averaging 1.39% above inflation (DIEESE, 2015, p. 2).

In addition to strikes by teachers, civil servants, and public transport workers in the public sphere, there was also a notable increase in activity by workers in the private sector, reaching 54% of the total in 2013 (DIEESE, 2015). Many of these strikes occurred in the service sector and were staged by unskilled or semi-skilled workers, many of whom were outsourced, underpaid, subject to precarious work contracts, and lacked conventional labour rights. In addition to eight national strikes carried out by bank employees, there was notable activism by workers in the transport, steel, tourism, cleaning, private health, safety, education, and communication industries. The most significant increase in strikes in the public sector took place in municipalities where union activity expanded to categories different from those that unions had traditionally mobilized, namely the most precarious workers in the city and regional administrations. Overall, in both the private and the public sphere, it is possible to identify an expansion of strike activity from the 'centre to the periphery' of the union movement, involving increased mobilization by the urban precariat.[2]

The strike initiated by the street cleaners of the city of Rio de Janeiro on 1 March 2014, aptly illustrates the main trends of the mobilization of precarious workers. Faced with employer intransigence, an insensitive municipality, and inaction by their own union leadership, the largely black urban cleaning workers organized themselves from below and demanded an increase in their meal subsidy and a raise of their meagre monthly salary. After eight days on strike, they won an historic victory: the municipal government agreed to increase their salaries from BRL800 to BRL1100 (at the time about USD470)[3] and their meal subsidy from BRL12 to BRL20 (about USD8.50). It is hard to imagine a more precarious group of workers. Karl Marx created a category to examine this part of the working class: he called it the 'stagnant' population, one step up from pauperism, that has to survive such degrading conditions of work that its social reproduction falls to subnormal levels (Marx, 2013).

In addition to their personal courage and combat readiness, the street worker's ingenuity was decisive for their victory. Instinctively, they built what might be called a 'symbolic politics of work' (Chun, 2009). They overcame the powerful obstacles to self-organization by making their problems public and making visible their hidden lives, thereby pushing towards a rebalancing of the scales. Thus they chose Brazil's most important popular festival, the 2014 Rio Carnival, to draw attention to their invisible lives and their importance to public life by letting garbage accumulate in the streets. At the same time, they marked the strike by holding massive workers' assemblies in public places. The public space of Rio de Janeiro came to be rendered in orange, the colour of the uniforms worn by the street cleaners, making them visible in the urban landscape (Braga, 2017). In this manner, the strike became an unavoidable issue in the public domain, stripping away the mantle of social invisibility that lay on this group.

This strike cycle and the vicissitudes faced by Brazil's subaltern classes in their precarious way of life are two faces that revealed the limits and ambiguities inherent in the Lulista project. To understand the contradictions of this project involves analysing the limits of the precarious hegemony established by PT in the last 13 years.

Precarious hegemony: passive and active consent

Understood as a *mode of regulation* of class conflict, Lulismo as a hegemonic social relation was based on the articulation of two different, but complementary, forms of consent, the consequence of which was the construction of a decade of relative social peace in the country. On the one hand, there was the *passive consent* of the subaltern classes to the government project, implemented through the trade union bureaucracy that ensured modest but effective concessions to workers (Singer, 2012). On the other there was the *active consent* of the trade union bureaucracy, leaders of social movements and intellectuals allied to the PT who were incorporated into the government project. We analyse each in turn.

We begin with an analysis of the passive consent for the PT government. Through the *Bolsa Família* Programme (Family Fund, PBF), the income of the semi-rural subproletariat rose from extreme poverty to the official poverty line during the PT governments from 2003–2011. By 2010, the programme assisted over 12 million families, 25% of the entire Brazilian population (Bello, 2016; Bichir, 2010; Rego & Pinzani, 2014). The associated expansion of credit allowed many Brazilians to purchase a car or motorcycle and travel by aeroplane for the first time. In fact, between 2001 and 2012, the number of automobiles and motorcycles in the country increased by 138% and 339%, respectively, in contrast to a population increase of 11.8% (Observatório das Metrópoles, 2013).

During the same period, the urban precariat was kept content by above-inflation wage increases and an increased rate of new jobs. On the back of a favourable global export market, Brazil created more than 2.1 million jobs annually between 2003 and 2010 (Pochmann, 2012, p. 32). Workers that were organized in unions benefited from this booming labour market and achieved both increased pay and other benefits through successful collective bargaining.

Table 1. Real increase in collective bargaining in Brazil (2008–2013) (Krein & Teixeira, 2014, p. 225).

Year	2008	2009	2010	2011	2012	2013
Average real increase	0.92%	0.90%	1.70%	1.36%	1.98%	1.25%

At least until the presidential election of 2014, this combination of redistributive policies, the increased rate of creation of formal jobs, and popular access to credit promoted a slight deconcentration of income inequalities. In a country known worldwide for its social inequalities, this small advance was strong enough to secure the consent of the subaltern classes to Lulista politics. It is noteworthy that Lula was easily re-elected in the 2006 election and his massive popularity reached 87% by the end of his second term in 2010.

Despite these gains for workers, the party ceded hegemony to the 'ideologues of the financial markets', as stated by the distinguished historian of the PT, Lincoln Secco (2011, p. 202). This was supported by Alfredo Saad-Filho who has demonstrated that the PT maintained 'the macroeconomic policy "tripod" (inflation targeting, floating exchange rates, and fiscal surpluses) enforced since 1999'. (see also Mollo & Saad-Filho, 2006; Saad-Filho, 2015b, p. 1246). Interest rates were among the highest in the world, and almost half of the revenue was used to pay off the national debt (Fatorelli, 2011). According to International Monetary Fund statistics, Brazil's four largest banks earned profits in 2013 that surpassed the gross domestic products of 83 entire countries (Cury, 2014; Horch, 2015).

The apparent economic success of the PT governments, however, masked important structural weaknesses. While Brazil retains a comparatively complex industrial base, it is still largely dependent on agricultural and mineral exports that subject it to the caprices of world markets (Morais & Saad-Filho, 2012). Even though the period from 2003 to 2010 was marked by economic growth and an increase in formal employment opportunities, 44% of the economically active population still held informal jobs without legal contracts, social services, or pension benefits. And it is particularly striking that the two single largest occupational categories in formal employment in the country are domestic workers (7.2 million) and telemarketing workers in call centres (1.4 million) (Braga, 2012, p. 212), both occupations that are also exploitative in nature and highly precarious.

While the government proclaimed that it brought tens of millions of Brazilians into the 'middle class', the economist Marcio Pochmann, himself affiliated with the PT, has proven this to be a myth: in fact, the middle class had contracted because of a flagging basic industrial sector (hitherto the provider of relatively high salaries and benefits) and the enormous expansion of contract labour (Pochmann, 2014). The creation of precarious low-wage jobs, the government's concerted effort at developing the agro-industry, and its reliance on orthodox financial policies made Brazil particularly vulnerable to the downturn in the world economy after 2008.

Thus, while there were indeed modest reforms in the areas of education, health care, and housing during the Lula presidency, these fell short of the expectations of the Brazilian majority. Spending on health care and education was woefully insufficient in relation to most developed nations and even other BRICS (Brazil, Russia, India, China and South Africa) countries (Gragnolati, Lindelow, & Couttolenc, 2013). A new social housing programme, *Minha Casa, Minha Vida* (My House, My Life), launched by the Lula government in 2009, only minimally alleviated the severe housing deficit and the lack of affordable housing and basic water and sanitation services continued. Instead of making a significant impact in the working class, the housing programme turned out to be a massive boon for private developers who controlled the siting and design of housing projects.[4]

While 18 new public federal universities were created from 2003 to 2011, modestly advancing local economic development in the towns where they were located and offering more options for post-secondary education to working-class children (albeit proportionally fewer than those offered by the profitable state-subsidized private universities), they lacked infrastructure and facilities, including laboratories, libraries, essential equipment, and lecturers (Cassol, 2012). The PT's programme of grants and low-cost loans to pay for private post-secondary education was massively utilized by the Brazilian population, but it also created huge profits for education companies (now publicly traded on the São Paulo stock market) that 'pasteurize' the content of classes, offering what every specialist agrees are poor-quality courses (De Almeida, 2014).

All these measures ensured the PT government the *passive consent* of the working classes. This was complemented by means through which the PT government gained *active consent* too. It did so by ably combining the interests of the trade union bureaucracy, the leaders of social movements, and the intellectual middle class – in particular, intellectuals that had fought against the military dictatorship and now worked in the academy and in the communication sector. The active consent for Lulismo was thus located in the state apparatus. For this, the PT government absorbed thousands of union members in parliamentary advisory functions, positions in ministries and state companies; part of the trade union bureaucracy assumed strategic positions on the boards of large pension funds managed by the state as investment funds; and PT members and supporters were nominated to management positions in the three main national banks, the National Development Bank (BNDES), the Bank of Brazil, and the *Caixa Economica Federal* (D'Áraújo, 2007; see also Jardim, 2009).

In this manner, Lulista unionism became not only an active administrator of the bourgeois state, but a key actor in the arbitration of capitalist investment in the country. Since this political-administrative power did not assume the form of private ownership of capital, the privileged social position of the trade union bureaucracy became dependent on its control of the political apparatus. In order to reproduce this control, the unions had to accommodate both the interests of their historic allies – low- and mid-level supervisors and managers in the public sector – and those of their historic enemies – hostile bureaucratic layers and sectarian groups with corporatist interests – within the state apparatus.

The creation of this active support also involved, since the 1990s, the construction of alliances between the PT and centre-right parties at all levels of government in order to guarantee its electoral viability and to obtain majority support in federal, state, and municipal legislatives (Secco, 2011, pp. 199–250). Former enemies, including politicians closely associated with the brutal military dictatorship (for example, Delfim Netto, Jader Barbalho, Fernando Collor, and José Sarney), were recruited by the government as political allies, power brokers, and consultants. And yet, while these measures assured electoral success for a short period, the alliances estranged many party militants and limited the government in advancing its overall programme.

These alliances worked well during the 2003–2011 economic upturn in Brazil, but they left the status-quo intact. They did not challenge the ideological dominance of neo-liberalism to any significant extent, nor did they push for reforms of the warped political system that gave disproportionate power to corrupt clientelist politicians (Saad-Filho, 2015a). At the same time, the integration of conservative politicians in political alliances severely hindered the PT's declared support for advances in social rights. These allied parties included many evangelical Christians and former police officers who effectively hampered initiatives in LGBT, black, and children's rights and attempts to rein in the brutal actions of the Brazilian security forces that kill poor people with impunity (Denyer Willis, 2015). The active consent thus garnered for the Lulista programme contained within it some problematic (but as yet untested) tensions and contradictions.

The June days: challenging the hegemony

By 2013, the Lulista hegemony had achieved a notable success in reproducing both the passive consent of the masses and the active consent of the leadership of the unions and social movements. The success of the Lulista hegemony in absorbing and reproducing the principal political tensions of the country obscured alternative projects of government that were not limited to traditional parliamentary bargaining. As the main social force on the left wing of Brazil's political spectrum and controlling the primary unions and social movements, the PT governments could have opted for a politics of mobilizing the subaltern classes, pressuring parliament for progressive political measures. At the height of their popularity, PT governments had a 72% approval rating. If they had chosen a different means of financing electoral campaigns (so that they were not indebted to the business sector and the elite) and had had a different relationship with parliament, it is quite likely that the 2016 coup could have been avoided.

Yet the PT hegemony was challenged dramatically by the entrance onto the political scene of the young precariat in June 2013. The Free Fare Movement (MPL) of São Paulo – a far-left wing-oriented political organization – had organized its fourth demonstration against municipal transport fare hikes. The protest turned violent, though only one side, that of the state, was armed. The brutal repression of protesters by the Military Police (PM) was a response to calls for the immediate restoration of 'order' in the city by governor Geraldo Alckmin of the Brazilian Social Democracy Party (PSDB), Mayor Fernando Haddad (PT), and many other political leaders of the city, including all PT and Communist Party of Brazil (PCdoB) councillors (Judensnaider, Lima, Ortellado, & Pomar, 2013; Moraes et al., 2014; Purdy, 2017).

Those who had followed the mobilization of the young workers against excessive increases in the cost of public transport were well aware that this protest was not limited to São Paulo. There were countless mobilizations in 2013 in other large cities in the states of Rio Grande do Norte, Goiás, Rio de Janeiro, Minas Gerais, Federal District, Ceará, and Sao Paulo, building on demonstrations in the previous year in the southern cities of Florianopolis and Porto Alegre. In the various cities, the concern of the protestors was expanded by other local issues, thus demonstrating the presence of profound social distress caused by the incongruity inherent in Brazil's development model.

Over the 2000s, millions of young workers were absorbed into the formal labour market. In fact, over 60% of the jobs created during the Lula and Rousseff governments were occupied by young people between 18 and 24 years of age. However, 94% of these jobs only paid up to BRL1000 a month (approx. USD450) (Pochmann, 2014). While the federal government significantly increased social spending, it proportionately decreased allocations to health and education. It invested untold billions in new World Cup stadiums, but underinvested in resources for urban mobility (Barros,

2014). Through the decade, conditions in the urban transport service declined, not only in terms of frequency and extent, but also in terms of quality and affordability. As Saad-Filho argues, 'rapidly rising incomes at the bottom of the pyramid and rising auto sales have not been accompanied by improvements in infrastructure, leading to an overall deterioration in the quality of urban life' (2013, p. 661). Long journeys on overcrowded, uncomfortable, and expensive vans, buses, trains, and subways were a daily fact of life for many Brazilians (Castelar, 2014).

From 2000 to 2012, bus fares in the country increased more than 67% above inflation (IPEA, 2013). The public transit systems of São Paulo and Rio de Janeiro were among the most expensive in the world (Dana & Siqueira, 2013). It was in this context, then, that the 'right to the city' became one of the underlying motivations of the 2013 protests, attracting widespread support from workers who lived in the far-flung peripheral areas of the large cities (Moraes et al., 2014). By prioritizing the financing of new cars for individuals, the federal government had encouraged profits for the auto manufacturers – and taxation for the state – at the expense of improvements in urban mobility and citizenship rights.

The violent police reaction to the demonstrations in São Paulo triggered a huge wave of social indignation around the country. In fact, June 2013 has gone down in the history of social rebellions in Brazil. Beginning on June 6 with a march of about 2000 people against the fare increases in São Paulo's public transport, the young MPL supporters could not have imagined that their actions would shake the country in an explosion similar only to the campaign for direct elections in 1984, during the military dictatorship.

Between June 19 and 20, 400 cities, including 22 state capitals, witnessed demonstrations and marches, bringing together approximately 8 million people (IBOPE, 2013). There were various reasons for this popular mobilization: an increasing weakness of the PT's development model that aimed at creating jobs and distributing income on the basis of the degrading use of cheap labour; a deepening international economic crisis that significantly slowed Brazil's economic growth rate; and the transformation of a more or less latent state of social dissatisfaction into widespread public anger by 2013.

The mobilizations against fare increases were not new and neither was the plethora of demands for the improvement of social services. The homeless workers' movement (MTST) had already begun to occupy empty buildings and mount large street demonstrations, employing many of the tactics historically used by the Landless Workers' Movement (MST) and redeployed by the MPL in 2013. One of the principal MTST leaders, Guilherme Boulos, explicitly acknowledged the June Days as an inspiration for intensifying its own campaign of occupations and street demonstrations in late 2013 and 2014 (Boulos, 2015). In many strikes and mobilizations throughout the 2000s, moreover, public sector workers had raised similar banners as the June 2013 protesters over the necessity to preserve and expand social programmes. It would also be negligent to ignore the impact of the Arab Spring protests and the Gezi Park mobilizations in Turkey that occurred at roughly the same time as the June Days.

The June Days had a significant influence on other movements. They boosted the profile of the *comitês populares da Copa* (people's cup committees) in cities hosting the 2014 World Cup, raising awareness of wasteful spending and corruption happening in the lead-up to the mega-event (Moraes et al., 2014). They also influenced a late-2013 strike by teachers in Rio de Janeiro, who in their tens-of-thousands mobilized a spectacular struggle of occupations and resistance despite intense police repression (Cocco, 2013; Mattos, 2015b).

Yet what we saw on the streets in June 2013 was a politically multiform movement, quite different from others in the recent history of the country. It was marked by a change over time in the profile of the protesters. It began when students and workers who use public transport on a daily level organized a range of demonstrations (in state capital cities such as São Paulo, Florianopolis, Porto Alegre, Vitória, and Salvador) under the banner of the MPL. They were joined from the outset by activists from far-left-wing parties such as the Party of Socialism and Liberty (PSOL), the United Socialist Workers' Party (PSTU), and the Brazilian Communist Party (PCB). After the violent police repression of the demonstrations in São Paulo, the protests extended to the urban peripheries, drawing in a plebeian mass of young people who blocked off several highways in São Paulo (Moraes et al., 2014).

Despite the incredulity expressed by Workers' Party leaders and cadres at the demonstrations, the June Days were not a surprise for researchers who had paid attention to the urban peripheries in the preceding months. Indeed, the protests were the rather predictable result of the limitations of Brazil's approach to development. This was demonstrated, for example, by research conducted in the Cidade Tiradentes district of São Paulo, which includes a large *favela* (shantytown) and one of the largest housing projects in Latin America (Cabanes, Georges, Rizek, & Telles, 2011).[5] The daily vicissitudes of the lives of the working families in this district, where 65% of the residents live on an average individual income of no more than BRL150 per month, revealed themselves abundantly in ethnographies of informal labour, drug trafficking, subcontracting, domestic work, illicit trade, police violence, illegal settlements, homelessness, and female-headed households (Cabanes et al., 2011). The June 2013 demonstrations transformed this myriad of private dramas into a fertile ground for public debate and confrontation.

It was also fed by the concerns that drove young women to reach higher levels of professional qualification, attempting to break out of the informal, domestic working worlds of their mothers into the formal world of telemarketing (Antunes & Braga, 2009). Despite the perception that telemarketing marked an occupational progress, the low wages (with salaries of no more than BRL1000 per month), harsh working conditions, high turnover rates and high-performance pressures led telemarketers to seek out unions for help. While this offered them temporary reprieve – access to payroll loans and the *Programa Universidade para Todos* (University for All Programme, Prouni) – by 2008 these measures were no longer sufficient and several strikes were called to demonstrate against low wages and poor working conditions (Braga, 2012).

The contradictions of Lulismo and the parliamentary coup[6]

Since 2008, Brazil has experienced a combination of an economic slowdown, strikes and popular mobilizations, and the erosion of the development model of which the redistributive limitations have become increasingly clear. During the economy's expansive cycle from 2003 to 2013, some social tensions were already evident, anticipating the current crisis. Despite the impressive increase in formal wage work during these years, 94% of the jobs created only paid up to 1.5 of the minimum monthly wage, approximately BRL1000. By 2013, the rate of jobs paying this amount had risen to 97.5%.[7] In addition, over the years an increasing number of accidents and deaths were occurring in these formal jobs and the turnover rate was rising, clearly indicating a deterioration in working conditions. The deepening economic crisis and shift towards an austerity policy during Rousseff's second government, in particular the fiscal adjustment programme with public spending cuts of BRL200 billion (USD85 billion), worsened the conditions of workers, prompting those who were members of unions to enter into strike action.

Although Rousseff's government was already faltering in 2014, it managed to win the second round of the elections that year with the support of the precarious proletariat. This support was conditional, however, on the continuation of formal employment opportunities, even if these were of low quality and poorly paid. And yet, since the election an economic contraction, driven by federal spending cuts, increased unemployment (which rose from 8.5% at the beginning of 2015 to 12% at the end of 2016 (Sales, 2017), hitting both the urban precariat and the organized working class.

The traditional middle class, in turn, some of whom were hitherto supportive of the PT and the main trade union federation, the Unified Workers' Central (CUT), moved towards a markedly right-wing economic agenda and politics. One important reason for this was the formalization of employment conditions for domestic workers, a form of labour widely employed by the middle class, which led to increased salaries at a time when the heated labour market raised the cost of services in general. In fact, the PT governments had effected a certain de-concentration of income among those who live from their jobs and this eventually produced certain effects which have had an impact on the middle class. Services provided by precarious workers to the middle class have seen their prices rise, so concierges, pedicures, manicures, and hairdressers, for example, are more expensive, and especially domestic employees. If one takes into account the tight labour markets, strategies to increase the minimum wage at more than the rate of inflation, which have a direct impact on domestic work, the cost of living for the middle class certainly increased significantly (Cagnin, Prates, de Freitas, & Novais, 2013). In addition, the increased buying power of workers led to their higher engagement in mass consumption. Workers began to occupy spaces, such as shopping malls and airports, that had previously been considered exclusive for the traditional middle classes. Lastly, the improving access to low-quality private universities for the children of workers meant increased competition for middle-class children for jobs that paid more than 1.5 the minimum monthly wage (Feres & Daflon, 2014).

Finally, the deepening economic crisis in 2014–2015 affected small and medium-sized businesses particularly sharply. Influenced by the reactionary and conservative corporate media in the country, the middle classes grew increasingly dissatisfied with the measures instituted by the government. When the Petrolão scandal broke, in which the state petroleum company Petrobas was linked to kickbacks and money laundering, the dissatisfaction of the traditional middle class and small and medium-sized businesses exploded into a huge wave of protest driven by a reactionary political agenda (Demier & Hoeveler, 2016; Purdy, 2015).

The collapse of Rousseff's support base in the National Congress was only the most visible face of a crisis that was rooted in the very social structure of a country that had suffered a deep recession for two years. The Brazilian development model, which had promised the creation of jobs for precarious workers and the evening out of income inequalities, was no longer able to guarantee corporate profits, let alone garner the consent of the subaltern classes.

Faced with a worsening international crisis, the main representatives of Brazilian business, with the private banks in the lead, began to demand that the federal government strengthen its austerity measures (Singer, 2016). In short, for large companies it was necessary to deepen the recessionary adjustment, increase unemployment, and contain the strike cycle in order to impose a series of unpopular reforms, such as cuts to social security and labour rights. This agenda flowed into the PT government's actions. The fiscal adjustment that Rousseff thus implemented early in her second mandate betrayed the expectations of the 55 million voters who had been seduced by her campaign promises of maintaining jobs, social programmes, and labour rights (Purdy, 2015).

For an analysis of the social roots of Brazil's political coup, we must consider two major dimensions of contemporary Brazilian society. The first necessarily relates to the question of the economy. After 10–12 years of relative growth, the country was in a recession. While it was able to uphold a low level of growth after the global economic crisis of 2008, the traditional instruments of economic policy, such as the adjustment of exchange rates, no longer worked in the recessionary situation after

2012. Second, when it became clear that a crisis was imminent, the Rousseff government began to institute reductions of social rights as a counter-measure to the economic woes. In particular, it attempted to reform unemployment insurance, loosen subcontracting regulations, and facilitate the privatization of public services (Rugitisky, 2015). It massively cut expenditure on health and education, a measure which affected the overwhelming majority of Brazilian workers who depended on state education and the public health system.

With these measures, the government implemented the agenda of employers, one that was based on privatization, the limitation of public expenditure, and the reduction of the alleged legal and financial burdens on businesses. The implementation of these measures undermined public resources in a period of recession (Rugitisky, 2015). With the naive attempt to restore the profitability of companies, the Rousseff government handled the public accounts and workers' savings irresponsibly. Implementing the employers' agenda over the last four years of the Rousseff governments led to a stalemate: the government's path of favouring accumulation by economic exploitation – even if it involved diminishing social policies – proved ineffective in restoring economic growth. In converse, it effectively deepened the economic crisis, amplifying and extending the economic recession (Bastos, 2015).

These financial problems precipitated the political crisis from mid-2015 onwards. It was marked by contradictory tendencies within the Rousseff government, the political and business elites, and the working class. On the one hand, sectors of the bourgeoisie aligned to the PT government understood the limits of a strategy of accumulation while also demanding that it be implemented more radically. On the other hand, the social base of the government was exerting enormous pressure for it to roll back these measures; and the Rousseff government, to some extent, depended on this base, even if the link between them weakened from 2013 onwards. This is the reason why the Rousseff government showed itself hesitant to call openly for items on the employers' agenda, such as diminishing labour rights, radically reforming unemployment insurance, allowing increased unemployment, providing more labour flexibility, or allowing informal work (Rodrigues, 2014).

It is no coincidence, then, that the Federation of Industries of São Paulo (FIESP) – having formerly supported the PT government by participating in a neo-developmentalist pact with the CUT and the government in 2011 – became one of the main organizations preparing the coup against Rousseff. It abandoned the government because it considered it too hesitant and ineffective, unable to further implement the social reforms it saw as necessary to re-establish the basis of private capitalist accumulation in the country. FIESP's proposals, presented during the impeachment campaign against Rousseff, put forward three principal employer demands for labour reform: the implementation the dominance of negotiated agreements over legislation; the right to implement flexible working days; and the ability to employ contract workers. Together with the abolition of the CLT (Consolidação das Leis Trabalhistas [Workers Law Consolidation]), these proposals, were they approved by Congress, would put an end to the degree of protections for workers that had existed in the Brazilian labour market since the 1930s. It would also threaten the Guarantee Fund for Employees (FGTS) and public transport and meal tickets.

Similar to FIESP, numerous employers abandoned Rousseff's government because they thought it too weak to push through measures that went against the PT's social basis. The idea of replacing her government with one led by the vice-president Michel Temer was seen to enable an administration that would unite all those who agreed on the necessity of implementing an agenda that would reduce social rights. This was the policy that the Brazilian bourgeoisie was betting on as solution to Brazil's economic crisis.

When the middle classes and small to medium-sized businesses took to the streets demanding the fall of the government, in direct response to the federal police's investigation of corruption at

Petrobas, their mobilization encouraged the political parties that had been defeated in the 2014 election to pursue an impeachment process against Rousseff to advance their neo-liberal agenda. The PSDB and the Brazilian Democratic Movement Party (PMDB) collaborated in the aim to repay all public debt to the lending banks, at the expense of spending on education, health, and social programmes.

It is critical to note that Rousseff's government was overthrown not because of what she gave (or did not give) to the lower classes, but for what she was not able to deliver to entrepreneurs and big business: namely, an even more radical fiscal adjustment that would require changing the Constitution, reforming social security, and withdrawing key labour protections. From the point of view of political logic, or the logic of social class domination, the Brazilian bourgeoisie was not prepared to accept a government that did not act to its satisfaction. The Rousseff government, in turn, was unable to follow the example of its predecessor government that, in the 2000s, had been able to absorb social movements, bureaucratically integrate social conflict within the state, and appease the sources of social dissatisfaction.

Final considerations

In the years leading up to 2014, Lula's first government had to absorb the anti-democratic rules of the Brazilian electoral system and break with existing presidential coalition schemes by buying political support in parliament. Despite this, it managed to establish a hegemony that lasted until 2014, built both on the passive consent of the masses and the active consent of leaders. During the expansive cycle of the economy in the first decade of the millennium, however, critical social tensions and contradictions began to manifest themselves, leading to a parliamentary, judicial, and media coup that led to the impeachment of Dilma Rousseff in August 2016.

With the deepening economic crisis and the sharp neo-liberal shift during Rousseff's second mandate, a post-Fordist and financialized accumulation regime intent on reducing social rights assumed dominance. From a developmental approach that flirted with peripheral Fordism, Brazil is now in a situation where the government has implemented austerity politics that were designed more or less single-handedly by the financial sector. Right-wing forces, with significant corporate media support, were organized and financed by organizations with clear class links – including conservative think tanks in the United States – with a concerted effort on impeaching Rousseff. In sum, the June Days' call for economic redistribution, with the clear demand for more state funds for social programmes and improvements in urban mobility, collided with the necessity of paying the interest that was accumulating on Brazil's public debt.

Yet, there is clearly an important relationship between the two waves of protest. After all, June 2013 was marked by the end of Lulismo and its social pacification project, and opened up a new political conjuncture. The June Days thus announced the arrival of a new era of class struggle in the country in which the mode of regulation that combined popular consent for the PT with consent by the leadership of the social movements was disintegrating.

The current economic crisis deepened the resentment felt by the traditional middle classes for the PT government, and strengthened anti-popular and conservative political currents. On the side of the popular masses, the austerity politics adopted by the Rousseff government alienated any popular support that had still remained for it. The crisis of Lulismo thus condensed the social contradictions that had built up over 13 years. Faced with a breakdown of the financial means for development, the Rousseff government decided to update the accumulation regime, prioritizing the strategy of social pillage.

Rousseff's submission to the pressures exerted by the financial markets should not come as a surprise. After all, the neo-liberal turn of her government had already begun early in her second mandate. The intensification of the tensions between a strident call for a regime of accumulation and the existing structures of regulation created an opportunity not just for a parliamentary coup, but, above all, for an adjustment of the PT development model that would reduce and weaken labour and social rights.

Some analysts aligned with Lulismo argue that the tensions between bourgeoisie mobilization and union and popular agitation, such as the 2013–2014 strikes and the June Days, eroded Rousseff's government and paved the way for the media, judicial, and parliamentary coup of 2016. This is an argument that is unsustainable if we compare the demands of June 2013 with the predominantly defensive union campaigns that took place in the same year. Instead, if we look at the new activism by precarious workers in the country, we can clearly identify the intransigent defence by precarious workers of social and labour rights. It was these redistributive efforts that the second Rousseff government openly attacked by implementing a politics of fiscal austerity and severe cutbacks after its electoral victory in 2014.

The recently elected government's betrayal of popular expectations, based on election promises, for a guarantee of employment and the inviolability of social right explains why the social basis of the PT government came to feel alienated. The intensified mobilization of the traditional middle classes, responding to calls made by the corporate media, occupied the space of the relative political vacuum that had been created by the June Days. From then on, the country experienced both a political and an economic crisis that had not been seen since the twin crises of the dictatorship and the model of peripheral Fordist development at the beginning of the 1980s.

The end of Lulismo signified the collapse of a hegemonic mode of regulation bound to the expansive cycle of the post-Fordist and financialized accumulation regime. The unfolding crisis of globalization in the country ended up diminishing the margins of concession to the subalterns, radicalizing the redistributive conflict, and precipitating a reactionary solution: a coup, the essential motive of which is exactly the deepening of neo-liberalism through a politics of social pillage, contrary to the expectations of Brazil's precarious workers.

Notes

1. By dynamics of and conflict around Brazilian distributive efforts, we understand, principally, the disputes over public funds by different social classes after the promulgation of the 1988 Constitution. The conflict pits the financial sector, which retains approximately half the government budget by way of interest and amortization payments on the national debt, against the majority of the Brazilian population, which depends on public spending for healthcare and education and on financial aid transfer programmes to the poorest in society (Druck & Filgueiras, 2007; Ferreira & Fragelli, 2012).
2. We define the urban precariat as a social amalgam formed by the most dominated and exploited sectors of the working class, as well as those fringes of the middle class facing proletarianization. It comprises workers that swing between formal and informal jobs in underpaid and degrading occupations or between deepening economic exploitation and full social exclusion (Braga, 2016).
3. All USD amounts given are historic amounts valid at the time of the events under discussion.
4. A side effect of this policy has been the formation of an intensely speculative bubble in the housing market, with house prices and rents rising at significantly higher rates than the increases in income and construction costs (Rolnik, 2015).
5. The larger aim of this research was to investigate the PT's micro-foundations and macro-hegemony.
6. Since there was no legal basis for Dilma Rousseff's impeachment, we refer to the process as a parliamentary coup. We use 'coup' here not in terms of its military connotations but to denote a break with existing

legislation and a reconfiguration of laws in terms of the interests of a particular interest group. In this sense, Brazil experienced a parliamentary coup in 2016, accomplished with the support of certain sectors of the judiciary, public prosecution offices, the federal police, business and media (Bianchi, 2016; Purdy, 2016).

7. These jobs were mostly occupied by women, young people and black people, thus workers who traditionally earn less and are more discriminated against in the labour market (Alegretti & Warth, 2014).

Disclosure statement

No potential conflict of interest was reported by the authors.

ORCID

Ruy Braga ⓘ http://orcid.org/0000-0002-8512-4306
Sean Purdy ⓘ http://orcid.org/0000-0001-6859-3525

References

Alegretti, L., & Warth, A. (2014, January 21). Criação de emprego em 2013 é a menor em 10 anos. *O Estado de S. Paulo*. Retrieved from http://economia.estadao.com.br/noticias/geral,criacao-de-emprego-em-2013-e-a-menor-em-10-anos,175913e

Antunes, R., & Braga, R. (eds.). (2009). *Infoproletários: Degradação real do trabalho virtual*. São Paulo: Boitempo.

Barros, M. L. (2014, March 11). Após greve, garis retornam ao trabalho com gostinho de vitória. *O Dia*. Retrieved from https://odia.ig.com.br/noticia/rio-de-janeiro/2014-03-11/apos-greve-garis-retornam-ao-trabalho-com-gostinho-de-vitoria.html

Bastos, P. P. Z. (2015). A Carta ao povo brasileiro, de Dilma Rousseff. *Revista Política Social e Desenvolvimento*, 3(1), 10–36.

Bastos, P. P. Z. (2017). Ascenção e crise do governo Dilma Rousseff e o golpe de 2016: Poder estrutural, contradição e ideologia. *Revista de Economia Contemporânea, 21*(2), 1–63.

Bello, C. A. (2016). Percepções sobre pobreza e Bolsa Família. In A. Singer & I. Loureiro (Eds.), *As contradições do lulismo: A que ponto chegamos?* (pp. 157–184). São Paulo: Boitempo.

Bianchi, A. (2016, March 26). O que é um golpe de estado? *Blog Junho*. Retrieved from http://blogjunho.com.br/o-que-e-um-golpe-de-estado/

Bichir, R. M. (2010). O Bolsa Família na berlinda? Os desafios atuais dos programas de transferência de renda. *Novos Estudos CEBRAP, 87*, 115–129.

Boulos, G. (2015). *De que lado você está? Reflexões sobre a conjuntura política e urbana no Brasil*. São Paulo: Boitempo.

Braga, R. (2012). *A política do precariado: Do populismo à hegemonia lulista*. São Paulo: Boitempo.

Braga, R. (2016). Terra em transe: O fim do lulismo e o retorno da luta de classes. In A. Singer & A. Loureiro (Eds.), *As contradições do lulismo: A que ponto chegamos?* (pp. 55–92). São Paulo: Boitempo.

Braga, R. (2017). *A rebeldia do precariado: Trabalho e neoliberalismo no Sul global*. São Paulo: Boitempo.

Cabanes, R., Georges, E., Rizek, C., & Telles, V. (eds.). (2011). *Saídas de emergência: Ganhar/perder a vida na periferia de São Paulo*. São Paulo: Boitempo.

Cagnin, R. F., Prates, D. M., de Freitas, M. C. P., & Novais, L. F. (2013). A gestão macroeconômica do governo Dilma (2011 e 2012). *Novos Estudos CEBRAP, 97*, 169–185.

Cassol, D. (2012, October 31). A universidade se universaliza? *Revista Desafios do Desenvolvimento (IPEA), 9* (74), Retrieved from http://www.ipea.gov.br/desafios/index.php?option=com_content&view=article&id= 2828:catid=28&Itemid=23

Castelar, A. (ed.). (2014). *Gargalos e soluções na infraestrutura de transporte.* Rio de Janeiro: FGV.

Chun, J. J. (2009). *Organizing at the margins: The symbolic politics of labor in South Korea and the United States.* Ithaca: Cornell University.

Cocco, G. (2013, December 7). 'O levante de junho: uma potentíssima bifurcação dentro da qual ainda estamos'. Entrevista especial com Giuseppe Cocco. *Instituto Humanitis Unisinos.* Retrieved from http://www. ihu.unisinos.br/entrevistas/526455-entrevista-especial-com-giuseppe-cocco

Cury, A. (2014, February 13). Lucro somado de 4 bancos brasileiros é maior que o PIB de 83 países. *O Globo.* Retrieved from http://folha.com/no1296233

Dana, S., & Siqueira, L. (2013, June 17). Análise: A tarifa de ônibus por aqui está entre as mais caras do mundo. *Folha de São Paulo.* Retrieved from http://www1.folha.uol.com.br/cotidiano/2013/06/1296233-analise-a-tarifa-de-onibus-por-aqui-esta-entre-as-mais-caras-do-mundo.shtml

D'Áraújo, M. C. (2007). *A elite dirigente do governo Lula.* São Paulo: Editora FGV.

De Almeida, W. M. (2014). *Prouni e o ensino superior privado lucrativo em São Paulo: Uma análise sociológica.* São Paulo: Musa/FAPESP.

Demier, F., & Hoeveler, R. (eds.). (2016). *A onda conservadora: Ensaios sobre os atuais tempos sombrios no Brasil.* Rio de Janeiro: Mauad X.

Denyer Willis, G. (2015). *The killing consensus: Police, organized crime, and the regulation of life and death in urban Brazil.* Berkeley: University of California Press.

DIEESE (Departamento Intersindical de Estatística e Estudos Socioeconômicos). (2015). Balanço das greves em 2013. *Estudos e Pesquisas, 79*, 1–44.

Druck, G., & Filgueiras, L. (2007). Política social focalizada e ajuste fiscal: As duas faces do governo Lula. *Revista Katálysis Florianópolis, 10*(1), 24–34.

Fatorelli, M. L. (2011, June 1). A inflação e a dívida pública. *Le Monde Diplomatique, 113*, 12–13.

Feres, J. Jnr., & Daflon, V. T. (2014). Políticas de igualdade racial no ensino superior. *Cadernos do Desenvolvimento Fluminense, 5*, 31–43.

Ferreira, P. C., & Fragelli, R. (2012, April 18). Desindustrialização e conflito distributivo. *Valor Econômico, 41*, 184–199.

Galvão, A. (2014). The Brazilian labor movement under PT governments. *Latin American Perspectives, 41*(5), 184–199.

Gragnolati, M., Lindelow, M., & Couttolenc, B. (2013). *Twenty years of health system reform in Brazil: An assessment of the Sistema Único de Saúde.* Washington: The World Bank.

Horch, D. (2015, August 13). In good times or bad, Brazil banks profit. *New York Times.*

IBOPE (Instituto Brasileiro de Opinião Pública e Estatística). (2013, June 18). 72% dos internautas estão de acordo com as manifestações públicas. *Ibope.* Retrieved from http://www.ibope.com.br/pt-br/noticias/ Paginas/72-dos-internautas-estao-de-acordo-com-as-manifestacoes-publicas.aspx

IPEA (Instituto de Pesquisa Econômica Aplicada). (2013, July 4). Tarifa de ônibus subiu 67 pontos percentuais acima da inflação. Retrieved from http://www.ipea.gov.br/portal/index.php?option=com_content&view= article&id=18865

Jardim, M. A. C. (2009). *Entre a solidariedade e o risco: Sindicatos e fundos de pensão em tempos de governo Lula.* São Paulo: Annablume.

Judensnaider, E., Lima, L., Ortellado, P., & Pomar, M. (2013). *Vinte centavos: A luta contra o aumento.* São Paulo: Veneta.

Krein, J. D., & Teixeira, M. O. (2014). As controvérsias das negociações coletivas nos anos 2000 no Brasil. In R. Véras de Oliveira, M. Aparecida Bridi & M. Ferraz (Eds.), *O sindicalismo na era Lula: Paradoxos, perspectivas e olhares* (pp. 213–246). Belo Horizonte: Fino Traço Editora.

Marx, K. (2013). *O capital. Crítica da economia política. Livro I: O processo de produção do capital.* São Paulo: Boitempo.

Mattos, M. B. (2015a, July 2). Junho e nós: das jornadas de 2013 ao quadro atual. Blog Junho. Retrieved from http://blogjunho.com.br/junho-e-nos-das-jornadas-de-2013-ao-quadro-atual/

Mattos, M. B. (2015b). New and old forms of social movements: A discussion from Brazil. *Critique, 43*(3–4), 485–499.

Mollo, M. L. R., & Saad-Filho, A. (2006). Neoliberal economic policies in Brazil (1994–2005): Cardoso, Lula and the need for a democratic alternative. *New Political Economy, 11*(1), 99–123.

Moraes, A., Gutiérrez, B., Parra, H., Albuquerque, H., Tible, J., & Schavelzon, S. (2014). *Junho: Potência das ruas e das redes*. São Paulo: Friedrich Ebert Stiftung.

Morais, L., & Saad-Filho, A. (2012). Neo-developmentalism and the challenges of economic policy-making under Dilma Rousseff. *Critical Sociology, 38*(6), 789–798.

Observatório das Metrópoles. (2013). *Evolução da frota de automóveis e motos no Brasil 2001–2012 (Relatório 2013)*. Rio de Janeiro: Instituto Nacional de Ciência e Tecnologia.

Pochmann, M. (2012). *Nova classe média? O trabalho na base da pirâmide salarial brasileira*. São Paulo: Boitempo.

Pochmann, M. (2014). *O mito da grande classe média: Capitalismo e estrutura social*. São Paulo: Boitempo.

Purdy, S. (2015, May 10). Rousseff and the right. *Jacobin*. Retrieved from https://www.jacobinmag.com/2015/10/dilma-rousseff-impeachment-pt-petrobas-brazil/

Purdy, S. (2016, August 12). Coups in Brazil: What's in a name? *ActiveHistory.ca*. Retrieved from http://activehistory.ca/2016/08/coups-in-brazil-whats-in-a-name/#comments

Purdy, S. (2017, April 19). Brazil's June days of 2013: Mass protest, class, and the left. *Latin American Perspectives*. doi:10.1177/0094582X17699905

Rego, W. L., & Pinzani, A. (2014). *Vozes do Bolsa Família: Autonomia, dinheiro e cidadania*. São Paulo: Editora Unesp.

Rodrigues, F. (2014). Presidente da FIESP fala em 'flexibilizar a lei trabalhista'. *Folha de S. Paulo*.

Rolnik, R. (2015). *Guerra dos lugares: A colonização da terra e da moradia na era das finanças*. São Paulo: Boitempo.

Rugitisky, F. (2015, May 8). Do ensaio desenvolvimentista à austeridade: uma leitura Kaleckiana. *Carta Maior*. Retrieved from https://www.cartamaior.com.br/?/Editoria/Economia/Do-Ensaio-Desenvolvimentista-a-austeridade-uma-leitura-Kaleckiana/7/33448

Saad-Filho, A. (2013). Mass protests under 'left neoliberalism': Brazil, June–July 2013. *Critical Sociology, 39*(5), 657–669.

Saad-Filho, A. (2015a, March 30). Brazil: The debacle of the PT. *Monthly Review Zine*. Retrieved from http://mrzine.monthlyreview.org/2015/sf300315.html

Saad-Filho, A. (2015b). Social policy for neoliberalism: The Bolsa Família programme in Brazil. *Development and Change, 46*(6), 1227–1252.

Sales, R. (2017, January 31). Desemprego no Brasil atinge a maior taxa desde 2012. *Valor Econômico*. Retrieved from http://www.valor.com.br/brasil/4853298/desemprego-no-brasil-atinge-maior-taxa-desde-2012

Secco, L. (2011). *História do PT*. São Paulo: Ateliê Editorial.

Singer, A. (2012). *Os sentidos do lulismo: Reforma gradual e pacto conservador*. São Paulo: Companhia das Letras.

Singer, A. (2016). A (falta de) base política para o ensaio desenvolvimentista. In A. Singer & I. Loureiro (Eds.), *As contradições do lulismo: A que ponto chegamos?* (pp. 21–54). São Paulo: Boitempo.

Singer, A. (2017a, September 2). Fisiologia pode brecar fetichismo da Lava Jato. *Folha de S. Paulo*. Retrieved from http://www1.folha.uol.com.br/colunas/andresinger/2017/09/1915245-fisiologia-pode-brecar-fetichismo-da-lava-jato.shtml

Singer, A. (2017b, December 23). Investigações mostram gritante desequilíbrio. *Folha de S. Paulo*. http://www1.folha.uol.com.br/colunas/andresinger/2017/12/1945624-investigacoes-mostram-gritante-desequilibrio.shtml

Neo-development of underdevelopment: Brazil and the political economy of South American integration under the Workers' Party

Fabio Luis Barbosa dos Santos

ABSTRACT

This article critically assesses Brazil's role in the South American regional integration process. My hypothesis is that despite the rhetoric of Brazil's Workers' Party (PT) governments about a 'new developmentalism' project to support 'post-neoliberal' regional integration, the structural continuities imposed by neoliberal macroeconomic policies have constrained all possibilities of overcoming underdevelopment. In the realm of regional integration, the driving force has been the internationalization of oligopolic Brazilian business in a process that promised Brazil a leadership role in the subcontinent. This frame has fostered business based on the overexploitation of labour and the destruction of the environment, enforcing trends that deepen the structures of economic dependency and social conflict. The political outcome of that process is that the PT has contributed to contain social pressures, both in the domestic and in regional contexts, as Brazil has played a moderating role in South America's so-called progressive wave.

1. Introduction

The election in 2002 of Lula da Silva, leader of the Workers' Party (PT), as president of Brazil raised high hopes, both nationally and worldwide. Forged in the final years of the military dictatorship (1964–1985), the PT was conceived of as a political front for left wing militancy, including labour unionists, activists of grassroots movements, liberation theologians and radical intellectuals. A major political force in the country and one of the largest leftist parties in the world, the PT stood in three national elections before its first electoral triumph in 2002.

Over time, however, the party has changed: its discourse has softened and its political practices have moved closer to conventional politics. Significantly, at the peak of his presidential campaign, while the PT led the polls and the threat of capital flight loomed, Lula released a 'Letter to Brazilians' that confirmed the party's commitment to the macroeconomics of neoliberalism. This commitment was duly enforced during the course of PT's four consecutive mandates. Despite significantly lowering its promises of what changes were to be expected from its governments, the PT managed to have its candidate elected in four successive presidential elections. However, as macroeconomic continuities prevailed, restraining change in every realm, the arguments to convince the left to vote for the PT shrunk considerably.

One of the last arguments in support of the PT, before being forced to vote for the least worst candidate, was its international policy, widely perceived as progressive, particularly in its immediate surroundings. During the period of the PT governments, Brazil vindicated its role as a regional leader by promoting a number of initiatives towards South American integration, such as the founding of the Union of South American Nations (UNASUR) in 2008. The friendly relations that Brazil sustained with leaders such as Hugo Chávez and Fidel Castro, particularly under Lula's presidencies (2003–2010), could be seen as further evidence that Brazil intended to defy the historical leadership of the United States (US) in the subcontinent.

This article challenges this view and argues that it is necessary to critically assess the role that Brazil played in the regional integration process. My hypothesis is that despite PT rhetoric about a 'new developmentalism' project in support of 'post-neoliberal' regional integration, the structural continuities imposed by neoliberal macroeconomic policies constrained the possibilities for substantive change. The driving force behind regional integration has been the internationalization of Brazilian oligopolic business, working within a frame of the prevailing international division of labour, in a process that promised Brazil a leadership role in the subcontinent. This rationality fostered business that was based on the overexploitation of labour and the destruction of the environment, enforcing trends that deepened the structures of economic dependency. The political outcome of this process is that the PT has collaborated to contain social pressures, both in domestic and in regional contexts, as Brazil has played a moderating role in South America's so-called progressive wave.

The argument develops in the following steps. First, the article places the election of Lula da Silva in a regional perspective, while discussing the broad outlines of the socio-economic policies adopted by PT federal administrations since 2003. In connection with this, Sections 3 and 4 examine the economic rationality behind the regional integration project, which has been deepened since then. In particular, I analyse the transnational dimensions of the accumulation strategies pursued by the Brazilian state under the so-called 'national champions' policy and the political premises underpinning this project. Social conflict and popular mobilization arising from these strategies are examined in Section 5. The final sections focus on the recent context in which economic pressures, brought about by the fall in international commodity prices, converged with corruption scandals to unleash a major political and economic crisis. This further exposed the contradictions of this project and ultimately led to the impeachment of Dilma Rouseff in 2016.

2. Neo-developmentalism and post-neoliberal regionalism

Lula da Silva's election as president of Brazil in 2002 should be seen in the broader context of a regional reaction against neoliberalism. The adoption of neoliberal policies in the preceding years had worn out the traditional political parties, while the precarization of labour had weakened historical worker organizations. Instead, new social movements moved into the foreground. As an outcome of these intertwined processes, new political leaders reached the presidency, such as Hugo Chávez in Venezuela (1998), Evo Morales in Bolivia (2006), Rafael Correa in Ecuador (2007), and Fernando Lugo in Paraguay (2008). Candidates that had historically identified with the political left were also elected, such as Tabaré Vazquez in Uruguay (2005) and Lula himself. In Chile, socialists took the lead under *Concertación* when Ricardo Lagos was elected (2000). Only Peru and Colombia, which began the 1990s under the spectrum of guerrilla war, leaned to the right.

The common ground shared by the South American governments that identified with this progressive wave was their asserted intention to confront neoliberalism. In Brazil, despite a macroeconomic regime grounded in the standard neoliberal triad of a target inflation rate, a target primary surplus to reduce the debt-to-GDP ratio, and a floating exchange rate, the political economy pursued by the PT governments (2003–2016) has been described as one of 'neo-developmentalism'. This label was commonly applied during Lula's second presidency (2007–2010), as a growth surge spurred on by higher commodity prices in the context of Chinese expansion combined with various state-led policies to foster national capital enterprises. Indeed, after decades of stagnation there was a slow recovery of the purchasing power of salaries; a slight improvement in the distribution of wealth; the reduction of extreme poverty through cash transfer policies; an expansion of consumption accompanied by abundant credit; and a lowering of unemployment rates. In addition, it was assumed that the country was sailing unharmed past the world economic crisis of 2008. These perceptions gave credit to the claim that Brazil was undergoing a new developmental period that could be compared to the national developmentalism that Brazil and parts of the Third World had experienced after the Second World War. Therefore, the neologism seemed appropriate.

The common thread between the diverse expressions of the neo-developmentalist discourse was the belief that the country should follow an alternative path between the financialization that typifies neoliberalism and the nationalism associated with developmentalism (Mercadante, 2010; Sicsú, de Paula, & Michel, 2005). According to this perspective, production should be prioritized over rentier capitalism, while taking pains to avoid inflation, fiscal populism, nationalism and other trends identified with national developmentalism. The challenge faced by neo-developmentalism was to reconcile what was believed to be the positive aspects of neoliberalism – such as its commitment to currency stability, fiscal austerity, a drive to international competitiveness, non-discrimination of international capital – with positive aspects of the old developmentalism – such as commitment to economic growth, industrialization, a regulatory state and social sensitivity (Sampaio, 2012).

The theoretical and practical contradictions of this incompatible association between what are supposed to be the beneficial aspects of neoliberalism with those of the old developmentalism have been analysed by several works (Fiori, 2011; Paulani, 2008; Sampaio, 2012). On the whole, the neo-developmentalist pastiche ignores the internal connections that articulate each proposition, besides disregarding the historical conditions that have given ground, in the past, to the national developmentalism utopia as a path to humanize peripheral capitalism.

The neo-developmentalist statement is matched in the field of international relations by the proposition that the PT governments implemented a 'developmentalist regionalism' or a 'post-neoliberal regionalism'. The common assumption underpinning both is that neoliberalism was left behind, or at least that Brazil was in the process of doing so, through policies that reshaped developmentalism ideals for the twenty-first century (Sader, 2010). Just as it happened in the economy, this understanding prevailed among PT supporters.

The former Brazilian Minister of Foreign Affairs, Celso Amorim, set the tone for this discourse early in Lula's presidency. In a well-known article, he presented the core ideas that should drive the country's foreign policy under PT: the deepening of regional integration, referred to a 'keen consciousness of the interdependency of Brazil and our South American neighbours'; the diversification of strategic partnerships with Asian and African countries, while preserving friendly relations with developed countries; and, last, the engagement with social issues, particularly through 'an international action to fight hunger and poverty' (Amorim, 2004).

Since then, Lula's foreign policy has been analysed and described sympathetically as a 'seasoned development' by Visentini (2008); as 'autonomy through diversification' by Vigevani and Cepaluni (2007); and as 'combined axes' by Pecequilo (2008). One of the most influential interpretations was proposed by Amado Cervo (2003). Endorsing Amorim's aim of combining 'the promotion of commercial liberalization and social justice', Cervo perceived a paradigm shift from what had been a 'normal state' to the so-called 'logistic state' in the present: 'The ideology beneath the paradigm of the logistic state associates liberalism as an external element, to Brazilian developmentalism as an internal element. It brings together the classical doctrine of capitalism with Latin American structuralism' (Cervo, 2003, 21).[1]

Despite acknowledging continuities between Fernando Henrique Cardoso's presidencies (from 1995 to 2002) and those of Lula, these scholars are clear that there was a major change in Brazilian foreign policy between the two men, though they analyse this change from different perspectives. In order to sustain that claim, the project pursued under PT has been referred to a shift in Brazilian diplomacy, in tune with the rise of progressive governments of different ideological hues. It has been argued that the Brazilian state dropped its focus on multilateral commercial liberalization, described by the Economic Commission of Latin America as 'open regionalism' (Comissão Econômica para a América Latina e o Caribe, 1994), in order to pursue a strategy of South American integration as a path to the region's sovereign insertion in the world order, referred as 'post-neoliberal regionalism' (Serbin, Martínez, & Ramanzani Júnior, 2012). It is held that such a trend would promote 'the physical integration between the countries' hinterlands, as a key step towards the integration of production chains of providers and producers, aiming at the formation of economies of scale and the integration of South American societies themselves' (Teixeira & Neto, 2012, p. 32).

On the other hand, the far right denounced PT foreign policy for various reasons: first, for what it saw as its links to an anachronistic Third World discourse; second, for its support of regimes considered to be authoritarian; and, last, for what it claimed was an infantile expression of anti-Americanism (Gonçalves, 2013, p. 106; Lampréia, 2013). The common ground between these critical views and the sympathetic approach is that both acknowledge that the Lula government introduced a major shift and both welcome the ongoing infrastructural integration, even if this is criticized by the right as being too slow.

3. South American integration and open regionalism

It is widely recognized, however, that it was in fact during the last years of the Cardoso presidency (and not only during the Lula one) that the perception emerged on the need to shift Brazil's foreign policy towards prioritizing South America. In 1994, the implementation of a free trade agreement between the US, Canada and Mexico made it clear that Mexico's policies would forthwith be tied even more tightly to the North American power. In the same year, Bill Clinton proposed the launching of a Free Trade Area of the Americas (FTAA) at the first summit of the Americas in Miami. In this context, Brazilian diplomacy pictured South America as the geographical reference for a regional leadership project, and it began to work on bringing the South American Common Market (Mercosur) closer to the Andean Community of Nations as part of a strategy to build an alternative economic bloc in the subcontinent (A Fundação Alexandre de Gusmão, 2012).

It was in this context that the Initiative for the Integration of the Regional Infrastructure of South America (IIRSA) emerged during Cardoso's presidency. Conceived as an institutional tool to coordinate a development plan involving all 12 South American countries, its goal was to set a common agenda of infrastructure projects in the areas of transportation, energy and telecommunication. If all on the initial list of more than 500 projects were to be carried out, the physiognomy of the subcontinent would be considerably modified, with a major social and environmental impact (Initiative for the Integration of the Regional Infrastructure of South America, 2011).

Proposed at a South American presidential summit in Brasilia in 2000, IIRSA was conceived as the infrastructural dimension of an integration project referred to the above mentioned open regionalism. The project portfolio was designed by the Inter-American Development Bank (IDB), which divided the subcontinent into 10 axes of integration and development with the aim of establishing new corridors for commodity export. From a geographic standpoint, the goal was to overcome the two 'natural' obstacles to subcontinental integration, the Andes mountain chain and the Amazon rainforest, in order to enhance connections between the Atlantic and the Pacific coast in the wider context of a displacement of the dynamic core of the global economy towards the East.

Following IIRSA's launch, Mexican president Vicente Fox announced the Puebla-Panama Plan (later renamed the Mesoamerica Project) with a similar aim, albeit on a smaller scale. Indeed, it was supported by the same multilateral financial institutions, particularly the IDB. Under these circumstances, IIRSA was construed as the infrastructural counterpart of the FTAA, which was widely repudiated by popular movements on the continent.

The FTAA was not implemented due to widespread popular resistance, though the Latin American bourgeoisie did not consent to it either. Equally important, it was not universally agreed on in the US itself, and George W. Bush's administration was unable to secure the fast track authority to speed up negotiations (Teixeira, 2011). The failure of the initiative became evident at the Fourth Summit of the Americas in Mar del Plata in 2005, a point that has been celebrated as a victory for Latin American progressive governments.

On the other hand, IIRSA was appropriated by the regional integration project of Lula's government. When UNASUR was created in 2008, its Council of Infrastructure and Planning (COSIPLAN) took up this initiative. It led to a curious paradox in which an initiative conceived of under the premise of open regionalism became the backbone of an organization that typifies 'post-neoliberal regionalism', its very antithesis. It was argued that it was both possible and desirable to incorporate the initiative's technical understructure, as Brazilian leadership under the aegis of UNASUR would grant a different political perspective to it. To fully understand this paradoxical outcome – as the FTAA was discarded and IIRSA was appropriated by UNASUR – it is necessary to analyse the economic motivations behind the Brazilian integration project.

4. The political economy of developmentalist regionalism

The economic drive underlying the PT governments' regional integration policy was its strategy to support the internationalization of large Brazilian corporations and companies headquartered in the country, which it saw as carriers of national capitalist development: its 'national champions' policy. This support manifested itself mainly through the entrepreneurial diplomacy engaged in by Itamaraty, the Brazilian ministry of foreign affairs, and through the credit policy adopted by the Brazilian Development Bank (BNDES).[2] In 2003, the bank modified its regulations in order to offer a new, special credit line to stimulate the international expansion of corporations that exported Brazilian goods and services. Loans granted for that purpose through the BNDES Exim programme jumped from US$42 million in 2003 to US$1.26 billion in 2009, an increase of roughly 3000% (Rodrigues, 2009). In 2010, total BNDES credit reached US$96.32 billion, which was 3.3 times higher than the US$28.6 billion that the World Bank lent in that year and a lot more than the US$11.4 billion mobilized by the IDB. That expansion was directly related to the growth of the Brazilian economy fostered by the commodities boom. In that year, the growth rate of Brazil's GDP was 7.5%, making the country the world's seventh largest economy (Leopoldo, 2011).

Besides earning a return on its operations, the BNDES's main source of funding was the at least 40% high compulsory saving that companies were required to make in terms of the Social Integration Programme and the Public Employee Savings Programme (PASEP),[3] as stipulated by Article 239 of the Federal Constitution of 1988. The Brazilian National Treasury channelled further resources to the BNDES. Indeed, the National Treasury's participation in the BNDES increased from BRL 3.8

billion (or 3.4% of the total) in 2001 to BRL 450 billion (or 54% of the total) in 2014. Some analysts argue that this was conceived also as an anti-cyclical measure to lessen the effects of the 2008 world economic crisis (Pinto & Reis, 2017).

However, these policies mean that BNDES loans were subsidized. The reason is that the bank charged so-called Long-Term Interest Rates (TJLP) for most of the loans it granted, at a much lower and less volatile rate than the Brazilian federal repo rate (SELIC) – in a context where the country's real interest rate at 5.64% in early 2015 was among the highest in the world, compared to 4.10% in China, 3.60% in Russia, 2.75% in India, and 0.45% in South Africa. By the end of Lula's presidency in 2010, when government bonds were based on the SELIC rate of 11.75%, the BNDES charged an interest rate of 6% on its TJLP loans. It has been calculated, for example, that the fiscal cost of the BNDES loans granted in 2009 was around 85% of the amount that the government channelled to its 'Bolsa-Família' scheme, Brazil's principal cash transfer programme, during the same period (Rodrigues, 2009). It goes without saying that the difference between government borrowing costs and the subsidized TJLP had a significant fiscal impact, affecting the gross national debt.

These policies benefitted the highly concentrated and oligopolized sectors of the Brazilian economy, which often operate as an extension of business chains that are dominated by transnational corporations, notably in the fields of civil construction and commodity exportation, involving products such as soybeans, ethanol, minerals, meat, oil, among others.[4] The government justified its focus, arguing that these were the sectors of the Brazilian economy that were better prepared to face international competition. Another way in which the PT governments supported the internationalization process was through BNDES Participações S.A. (BNDESPar), BNDES's investment arm that capitalized companies through the acquisition of stock or debentures. In 2009, BNDESPar had equity participation in 22 multinationals headquartered in Brazil, and its investments amounted to BRL 92.8 billion, corresponding to 4% of the Brazilian stock market (Tautz, Siston, Pinto, & Badin, 2010, p. 261). In 2012, 89% of BNDESPar equity was concentrated in the oil, mining, paper and cellulose, energy, and meatpacking sectors (Garzon, 2013).

The BNDES collaborated to intensify capital concentration in several sectors of the Brazilian economy. This trend was accelerated during the 2008 economic crisis and reached its peak in 2010 when there were over 700 merger and acquisition operations involving Brazilian firms. For example, the BNDES provided BRL 6 billions to the JBS group for acquisitions in Brazil and overseas, creating the largest meatpacker in the world; it allocated BRL 2.4 billion for the Votorantim Celulose e Papel acquisition of Aracruz Celulose, which resulted in one of the world's largest cellulose producers, Fibria; and it granted over BRL1.5 billion for the merger between Sadia and Perdigão which created the Brasil Foods group, the largest chicken processor and exporter in the world.

JBS is a case in point: two years after the BNDES provided the financing for it to acquire Swift Armour, the largest meatpacker in Argentina, BNDESPar injected BRL 4.5 billion into the JBS-Friboi conglomerate for it to acquire two companies in the US, Swift&Co. and Pilgrim's Pride Corp. In 2008, BNDESPar held 20% of JBS's stocks, its participation reaching 35% in subsequent years as a way to eliminate debts and debentures. Until 2010, the BNDES invested over BRL 7.5 billion in JBS (Rodrigues, 2010). In 2017, the company came to be at the centre of corruption scandals that shook Temer's presidency.

JBS's path also illustrates the move of Brazilian capital towards Argentina, a country that went through a severe economic crisis at the turn of the millennium, culminating in mass demonstrations that, by the end of 2001, had deposed five presidents in a row. In 2003, the Brazilian state-owned oil company Petrobras gained control of the Argentinian Perez Companc conglomerate, and has since increased its presence in the country; in 2005, the Brazilian Camargo Corrêa bought Loma Negra,

Argentina's largest cement producer; in 2010, the Brazilian mining company Vale acquired the Argentine assets of the British-Australian corporation Rio Tinto (Rio Colorado potassium project); the Brazilian Votorantim, which already had stocks of steel company AcerBrag, bought 50% of Cementos Avellaneda; and, in the service sector, Banco do Brasil acquired the Argentinian Banco Patagonia (Fundação Dom Cabral, 2010; Luce, 2007).

However, the protagonist of international expansion by Brazilian business was civil construction, a sector that was built up under the Brazilian dictatorship (1964–1985) and that diversified its activities during the privatizations process of the 1990s, a process also supported by the BNDES (Campos, 2009). As a rule, these firms turned into diversified conglomerates with civil construction as one of their business interests. Under PT, the construction sector was domestically boosted through the Growth Acceleration Programme (PAC), so that some analysts interpret BNDES support for international expansion as an extension of this programme (Garzon, 2013). BNDES loans to Brazilian building contractors abroad increased from US$72,897 million in 2001 to US$937,084 million in 2010, a total of 1185%. During Lula's presidencies, more than US$10 billion were channelled to finance construction projects in the region, many of them connected to the IIRSA. These projects included the gas pipeline network (US$1.9 billion, by Odebrecht and Confab) and the Chaco aqueduct (US$180 million, by CNO, Techint, OAS, and Isoluc), both in Argentina; the San Ignacio de Moxos-Villa Tunari road (US$332 million, by OAS), the Hacia el Norte-Rurrenabaque-El-Chorro Project (US$199 million), and the Tarija-Bermejo road (US$179 million, by Queiroz Galvão), all three in Bolivia; the Santiago subway in Chile (US$209 million, by Alstom); a bridge over Tacutu river in Guiana (US$17.1 million); a second bridge over the Parana river in Paraguay (US$200 million); the Assis Brasil-Iñapari bridge in Peru (US$17.1 millions); the gas network of Montevideo, Uruguay (US$7 millions, by OAS); and in Venezuela, the Carcas subway (US$943 million, by Odebrecht) and the La Vueltosa hydroelectric plant (US$121 million, by Alstom). To that list should be added large hydroelectric power plants projected in Peru, Venezuela, Ecuador, Colombia, and the Dominican Republic by constructors Odebrecht, Camargo Correa and OAS (Garcia, 2012).

The Peruvian case in particular illustrates Brazil's regional strategy. In 2010, the respective presidents Alan García and Lula da Silva signed an agreement foreseeing the construction of five hydroelectric power plants in the Peruvian rainforest, which would export up to 80% of their production to Brazil. The first and principal among them was to be built in Inambari, a project that was granted to a consortium of three Brazilian firms, led by OAS. The thought behind the project was to unleash a process to bring the Peruvian economy closer to Brazil, as had happened in the 1990s when a gas pipeline made Brazil Bolivia's main commercial partner. According to the Brazilian view, tighter economic bonds would build up the material basis to further the political autonomy of the region, under Brazilian leadership.

However, the outcome of the Inambari proposal was a telling if unexpected statement on the nature of the Brazilian integration project: massive popular protests, motivated by the impending social and environmental impacts of the project, blocked the construction as Peruvians widely considered the project to be driven by foreign interests. The project stalled and, by late 2016, had still not been reactivated.

When not hampered by popular unrest, the rationality underlying the PT governments' project assumed that the internationalization of Brazilian corporations would serve as the material basis to consolidate the country's regional influence, in a process that would enhance its international status. Or to state it in diplomatic vocabulary, Brazil would become a 'global player'.

UNASUR was created in 2008 under that very perspective. It differed from the Bolivarian Alliance for the Peoples of Our America (ALBA) that had been led by Hugo Chávez with Cuban support since

2004 with the aim of advancing integration beyond a market-driven approach, aspiring to build a Latin American bloc counter-hegemonic to the US. Limited to South America, UNASUR has been marked since its inception by an attempt to find the smallest common denominator that could embrace governments of contrasting orientations, such as Venezuela and Colombia in that context.

Beyond accomplishments that are welcome but that have little structural impact, such as agreements to validate each other's university degrees or joint purchases of vaccines, it was expected that the creation of COSIPLAN as a branch of UNASUR would reverse IIRSA's original orientation. In economic terms, this was meant to reduce the role of the multilateral financial institutions that conceived the initiative, namely the IDB, the Andean Development Corporation (CAF) and the Financial Fund for the Development of the River Plate Basin (FONPLATA). The reason for this was that these institutions' strictly market-driven orientation was seen as a hindrance to infrastructure work that was necessary to promote integration but that would possibly not generate profits. The aim was to counterbalance them by other institutions, notably the BNDES. Simultaneously, proposals around a 'new regional financial architecture' gained ground, resulting in the Bank of the South in 2009, joined by Argentina, Bolivia, Brazil, Ecuador, Paraguay, Uruguay, and Venezuela.

From a Brazilian point of view, the integrationist trend generated business opportunities for its companies and strengthened the country's political role. Other countries in the region had two fundamental motivations to join in. On the one hand were those that wished to consolidate a political field alternative to the historical influence of the US. That had been the motivation behind many South American countries joining the ALBA initiative, whose radicalization potential was, however, neutralized by the Brazilian course of action. On the other hand were the countries who were doing business with Brazil as with any other country, such as Peru. Characteristically, Venezuela and Peru were the South American countries where the corporation that most distinctly symbolized PT-supported expansion, Odebrecht, had its strongest presence.

As a rule, the countries that intensified their commercial connections with Brazil were the ones with which PT administrations cultivated close political bonds, such as Argentina and Venezuela on the continent, and Cuba and the Dominican Republic in the Caribbean. Outside the continent, the link between political affinity and commercial connections seemed to be less relevant, considering for example the amount of business done with the dictatorship of Angola. But political allies could certainly make business smoother, and that was the motivation behind Brazil's active support of Ollanta Humala in Peru's election in 2011 – although the candidate turned his back on his Brazilian supporters once he was elected.

5. Developmentalist regionalism as experienced from below

Many of the projects supported by the BNDES have had a considerable social and environmental impact, besides benefitting firms that systematically disregarded labour legislation. For all those reasons the Bank's role has attracted popular opposition, both in Brazil and in the region more widely. In 2007, diverse social organizations and popular movements launched the BNDES Platform, which highlighted the contradictions between the bank's current function and the social concerns that should underlie its role as a public entity.

There are several examples of BNDES's controversial role. São João sugar mill, part of the Brenco group, received a BNDES loan of BRL 600 million and had BNDES-Par as its partner, but was charged by the Public Prosecutor's office with keeping 421 workers in a situation analogous to slavery. Meatpacker Bertin, which BNDESPar controlled to 27.5% after injecting BRL 2.5 billion (before it was acquired by JBS), was frequently denounced for keeping cattle in areas that had been illegally deforested. In yet another case, the Alcoa group received a BNDES loan of BRL 500 million to mine bauxite in Juriti municipality, even though it had not acquired a valid environmental permit, leading to harsh conflicts with the local population (Tautz et al., 2010).

Among these conflicts, two in particular came to the foreground and drew wide public attention. In 2011, a rebellion erupted among the 14,000 workers at Jirau dam in the state of Rondonia, one of the largest hydroelectric constructions in the country close to the Bolivian border. This conflict triggered rebellions in similar work environments around the country. These workplaces are often very remote, well removed from any presence of the state and from the eyes of the public opinion, and become sites of hyper-exploitation of the work force. The project promoter in Jirau is constructor Camargo Correa, associated with French transnational Suez and Eletrosul, and they were granted BRL 13.3 billion from BNDES. A second conflict that reached an international audience involved the construction of the massive Belo Monte hydroelectric power plant by Xingu River, in the heart of the Brazilian rainforest. Despite the doubtful urgency of the project, its uncertain economic viability and its extraordinary social and environmental impact, the construction is well on its way and has to date mobilized BRL 23 billion from the BNDES.

The role of the BNDES has been contested in the international arena as well. The most famous conflict involved work relations at the International Nickel Company of Canada (Inco), which the Brazilian giant Vale do Rio Doce acquired in 2006. Workers in Canada faced the new management with a long-term strike that raised solidarity among Vale workers worldwide.

In the South American sphere, there have been multiple controversies. In Ecuador, irregularities detected in the construction of the San Francisco hydroelectric power plant led to a conflict between Correa's government and Odebrecht, escalating to the brink of a diplomatic crisis in 2008. In Bolivia, conflicts triggered by plans to build a road through the Isiboro-Sécure Indigenous Territory and National Park (TIPNIS), at once a national park and an Indian reserve, are regarded as a turning point in the relationship between the Morales government and indigenous movements. The construction was granted to Brazilian OAS, backed by a US$332 million BNDES credit. Despite brutal state repression of the eighth indigenous march in 2011, work on the road had to come to a halt due to fierce popular resistance. A similar situation arose in the Peruvian rainforest where the construction of hydroelectric plants by Brazilian contractors was blocked by social protest.

6. Developmentalism and crisis

Though popular contestation eventually blocked a few specific projects, it did not threaten the overall dynamic of developmentalism. How, thus, has the regional integration process evolved? From the standpoint of Brazilian foreign policy, which envisaged capitalist development backed by the internationalization of Brazilian corporations and, as a consequence, aspired to a protagonist role for the Brazilian government in international politics, it can be asserted that this project moved on with relative success during the Lula years. The commodities boom sealed his unchallenged popularity, even allowing the election of a virtual unknown as his successor. At the same time, there were several signs that the country was playing a new role – from the leadership of the United Nations peace mission to Haiti from 2004 onwards, and the victorious campaign to host the FIFA World Cup of 2014 and the summer Olympic games of 2016 – even if these accomplishments were of questionable virtuousness.

During Rousseff's first presidency (2011–2015), the process was still moving forward, although at a more intermittent pace. While the new president did not prioritize the international agenda nor regional integration, there were signs that a crisis was developing. As growth stalled, resources chan-nelled to the BNDES by the National Treasury decreased. On the political front, although the mass rebellion that took place in June 2013[5] was not triggered by economic distress, it did convey a wide-spread social malaise. It was a clear sign that the social pacification produced by PT, where lower classes accessed marginal gains while business profited as ever, was loosing ground. As popular demands were frustrated again, political momentum shifted consistently to the right.

When Rouseff was re-elected in 2015, the PT project was threatened on several fronts. Essentially, the conditions that protected Brazil from the harshest impact of the 2008 economic crisis no longer pertained. On the contrary, there were signs that an economic recession was on the cards. At the same time, numerous scandals brought to light corruption networks that seemed ingrained in how the construction industry was doing business, in the country and abroad. The continuing exposure of corruption schemes undermined the government's reliability and popularity, as well as the political party that was at its head. But it also threatened the cornerstone of the capitalist pro-ject that the party and the government advocated. Strikingly, at this point the call for transparency in BNDES accounts was no longer made only by popular movements, but also by the political right that took to the streets to make its point.

Assailed on every front, PT cadres interpreted the interruptions faced by the contractors and Pet-robras as a threat to its national project. Unable to deny the corruption charges, they alleged that the country had always been run that way, but that this was surfacing only now because it was PT that was in the presidency. In parallel, the cadres suggested that the comprehensive investigations into corruption that were being pursued inhibited investments by contractors and other businesses, jeo-pardizing the economy as a whole. Through these tensions it became evident that the regional inte-gration project assumed the interests of contractors and the nation to be identical, regardless of the former's corrupt way of doing business, not to mention the precarious work conditions they maintained.

To recognize the fragility of the PT project does not imply endorsing the political use that anti-PT right made of the corruption investigations, nor the reprehensible irregularities incurred for that purpose. However, the scandals made it impossible to deny the corruption. And the conjunction between corruption and the economic crisis cowed the government into a defensive position. Never-theless, even before the crisis became acute, the BNDES gave signs that it was worried about the investigations rather than about its investments. In any case, budget cuts together with a decrease in revenues from the Workers' Support Fund (FAT), another traditional source of funding for the BNDES, were reducing the capital it had available for long-term loans.

At this point, several indicators began to raise questions about the efficacy of the 'national cham-pions' strategy. First, several companies that had received sizeable BNDES grants passed into foreign control. That was the case, for example, of the beverage conglomerate Ambev that merged with a Belgian corporation and transferred its headquarters to that country; it applied to the ethanol plant Santa Elisa, acquired by French LDC Dreyfuss soon after having received BNDES funding; and to EBX, which linked up with Chinese and Korean capital after having received large public credits for its projects; as well as to Alunorte and Alumar, sold by Vale to Norwegian Norsk Hydro. There were also situations like where Odebrecht became an autonomous operation in Peru: legally registered as a local enterprise, it was no longer supported by the BNDES, but neither did it contribute to the exportation of Brazilian products and services. Overall, there is no consistent evidence that services provided by the BNDES stimulated output growth. Instead, there is evidence that 'national champions' borrowed long-term funds either to reduce capital costs or even to benefit from interest rate arbitrage profits (Bonomo, Brito, & Martins, 2014).

The domestic difficulties faced by the PT governments were compounded by the difficulties changing IIRSA's orientation through COSIPLAN. The fundamental argument for incorporating IIRSA was that the projected infrastructure would contribute to endogenous growth, environmental sustainability and social inclusion. However, the banks that backed the initiative right from its inception (the IDB, the CAF and Fonplata) have retained the power to dictate IIRSA's project portfolio. Thus, these banks finance the feasibility studies that precede any project, and their representatives still dominate IIRSA's technical committee under COSIPLAN.

In addition, the proposition for a 'new regional financial architecture', designed to reduce the region's dependence on international funds, was blocked – by Brazil itself. Both Brazil's Federal Reserve Bank and the National Treasury consistently objected to using national reserves to finance investment. Instead, U.S. Treasury bonds with an interest rate of 1% per year were bought, despite the fact that, at home, BNDES was paying out the highest interest rates in the world. Although one of the main proposers of the new scheme, the Ecuadorian economist Pedro Paes, minimized the issue by arguing that the input of reserves would be preceded by numerous other measures, the perception prevailed that the Bank of the South did not take off for political reasons: the institution proposed an equal vote for every member, regardless of their financial input, which is a different structure from that of the International Monetary Fund (IMF). This approach did not interest the Brazilian leadership, however, a point that emphasizes the power motivations that underlie its regional integration discourse.

The fragility of the integrationist proposal is also attested to by COSIPLAN's project portfolio, to which the IIRSA is attached. In 2014, 477 of the IIRSA projects had a national range, 95 were binational, 5 were tri-national and only 2 were multinational, both in the area of telecommunication. The IIRSA report for that year stated that 89.1% of the projects and 66.5% of the foreseen investments were related to the transport sector, almost half of them for highways; energy projects amounted to 9.3% of the projects and 33.5% of the investments, and communications was less than 2% of the projects. These figures demonstrate the affinity between IIRSA and the core business of Brazilian contractors (Consejo Suramericano de Infraestructura y Planeamiento, 2014). Fifteen years after its launch, the IIRSA initiative has been making slow but consistent progress, yet without fulfilling the political role that the PT project had aspired to.

This observation problematizes the alleged nexus between neo-developmentalism and regional integration. Since the economic dimension of the process, basically consisting of construction driven business, was conceived under the frame of open regionalism precepts, the proposition of a developmentalist regionalism involved a double rhetorical maneuver, evoking a shift in a project that remains essentially the same. It associates the expansion of oligopolic Brazilian and transnational business with new developmentalism while, at the same time, identifying this new developmentalism with post-neoliberal integration. Under that perspective, interconnection is misconceived as integration, oligopolic interests as national interests, entrepreneurial diplomacy as South-South cooperation, and the internationalization of Brazilian business as post-neoliberal integration.

7. PT order in South America

Lastly, the management of regional conflicts in South America, which is the main virtue commonly ascribed to UNASUR, was mistaken as sovereignty. According to that interpretation, the creation of regional organizations outside of the purview of the US should be seen as a breakthrough towards a multipolar world, creating an opposition to superpower interests in the region. However, it should be highlighted that the US State Department did not hold the view that South American leadership crowded out its own presence. The then Secretary of State Condoleezza Rice openly supported Brazilian leadership when UNASUR came into being (Rice, 2008). Under the Obama administration, the sub-secretary of Political Affairs of the State Department Wendy Sherman emphasized: 'Today Brazil is a strategic partner in addressing global – not just hemispheric – issues of shared concern. And I want to be clear that the United States needs and welcomes Brazil's positive expanded role' (Sherman, 2012).

In fact, so far Brazilian intervention in regional affairs has not antagonized the US. Brazil interceded against coups that had U.S. sympathy in the recent political crisis in Honduras (2009) and in Paraguay (2012). Despite this, however, Brazil was unable to reverse the course of events in either situation, although it made a remarkable effort in the Paraguayan case.

The limits of PT's progressive stand becomes evident in its relations with the Bolivarian process in Venezuela. While Lula cultivated a close relation with Chávez, he also helped marginalize the latter's most innovative initiatives, such as ALBA, the Bank of the South, and Telesur, a multi-national state-funded television network. The Brazilian attitude empowered the moderate sectors within Bolivarianism, a dynamic political process whose evolution was permanently contested. In the process, Brazilian business in Venezuela multiplied.

As the PT leadership neutralized the most radical expression of the progressive wave which is currently fading, they have contributed to limit the scope of change during this period. In some situations there has been radical political change, as in Venezuela where the Puntofijo Pact[6] was wiped out after having given structure to the politics of this country for four decades; or in Bolivia, where the social segregation between indigenous and non-indigenous, reproducing a 'sociedad abigarrada' (multi-layered society),[7] was eroded by indigenous protagonism. However, it has not been possible to establish an alternative path to an economy based on commodity exports in any of these cases. This explains the unanimous adhesion to IIRSA, as all South American governments, including Venezuela, supported the expansion of export corridors. There is, therefore, a correspondence between the progressivist process in the region, where there has been political change tied to economic continuity, and the dynamic of regional integration, where UNASUR has emerged as a political novelty tied to IIRSA, which was framed under a logic that deepens the structures of economic dependency.

This ambiguity between political progressivism and economic conservatism has generated contradictory situations, as revealed during the crisis in Paraguay in 2012. Fernando Lugo's election in 2008 was the first political change in this country after six decades of rule by the National Republican Association–Colorado Party, which included the longest dictatorship in the region (1954–1989). Despite regarding Lugo as a regional ally, the PT administration interfered in Paraguay's situation by fostering credit lines and giving political support to so-called 'brasiguayos' (Brazilian businessmen who had settled in Paraguay) for the expansion of soybeans. This approach strengthened those political sectors in Paraguay that opposed any initiative to democratize access to land, as Lugo's campaign had promised. As a consequence, Lugo's political position was consistently undermined, until he was deposed in an illegitimate impeachment process that Brazilian diplomacy was powerless to stop (Dos Santos, 2014). This was an emblematic situation in which support for economic sectors that opposed political change finally made any change impossible. As a matter of fact, it can be argued that this ambiguous trend played a role in the process that led to the impeachment of Rouseff herself in 2016.

8. Concluding remarks: post-neoliberal regionalism and ideology

There is a correspondence between neo-developmentalism as an ideology of the political economy of the PT presidencies and developmentalist or post-neoliberal regionalism as its foreign policy ideology. In both cases, the political function is to differentiate themselves from preceding governments and their neoliberal orthodoxy. Nevertheless, no substantive changes took place on the ground in line with these claims, either in the macroeconomic policies set by the Plano Real stabilization plan of 1994, or in the regional integration process that had IIRSA as its backbone. In any case, these policies were based on conceptions that Celso Furtado (1974) has described as the myth of economic growth. Therefore, the spurious antagonism that the PT government created between itself and its predecessors restrained the scope of political debate in such a way that it elided the connections between economic growth and the articulation of dependency and inequity, which characterize underdevelopment. Propositions pointing to an alternative civilizatory path around 'sumak kawsay' (good living),[8] bolivarianism[9] or socialism did not have a space in this.

The economic debate in the country is restrained to microeconomics, revolving around the pace and intensity with which the neoliberal agenda should be executed. In the field of international relations, the relevance of the South in general and of South America in particular is disputed, in the light of a mercantile rationality. The proposition of a privileged sphere for the expansion of Brazilian business is weighed against the liabilities of the regional integration process. Vera Thorstensen, the former Brazilian assessor at the World Trade Organisation, described the counterpoint to the PT position with crude clarity when she said with reference to Mercosul: 'it is no use to marry the poor' (Drummond, 2014). On the other hand, IIRSA is supported by diplomats aligned with open regionalism. For instance, José Botafogo Gonçalves, who served under Cardoso, complained that the initiative 'has been disregarded, when it is perhaps the institution that can bring greatest dynamism to South American integration', while defending the practice of 'infrastructural diplomacy' (Gonçalves, 2013, p. 268).

Despite different approaches, there is a common basic aim: insert the Brazilian economy into the trends of contemporary capitalism as a commodity exporter, as a basis for transnational capital operations and as a platform for financial capital valorization. Contrary to the post-neoliberal rhetoric, the meaning of the integration process has not been modified since the election of progressive governments; rather, the mercantile interconnection of the subcontinent has been enhanced by the political affinity between these governments.

From this angle, the seeming contradiction of a conventional government adopting conservative social and economic policies but undertaking an innovative foreign policy to disturb US interests is resolved. As the ideological articulation between neo-developmentalism and post-neoliberal regionalism is unveiled, the role of the PT political economy for South America emerges not as what it would like to be, but as what it has been: the instrumentalization of regional integration for the purposes of the internationalization of Brazilian oligopolic business on the basis of the prevailing international division of labour, in a process that promised Brazil a political leadership role in the subcontinent, to be exercised at critical moments in line with the strict limits tolerated by the hegemonic power.

Notes

1. All Portuguese quotes have been translated by the author.
2. The National Bank for Economic and Social Development, by its full name, is a state-owned development bank that was founded in 1952.

3. The PIS and the PASEP are compulsory savings required from companies, established by Brazilian law in 1970.
4. Powerful Brazilian soybean growers in Paraguay, for example, get their inputs mostly from Monsanto. At the other end of this supply chain, the commercialization of the soybean is undertaken by corporations such as Cargill, Archer Daniels Midland and Bunge.
5. In June 2013, massive mobilizations occurred in over a hundred cities against fare hikes for public transport, poor-quality education and health care services, and the immense public investment in 'mega-events' such as the 2014 Football World Cup and the 2016 Olympic Games.
6. The Puntofijo Pact was an agreement between the leading political parties in Venezuela in 1958, which led to an alternation in the presidency between the two principal parties, the AD and Copei. The Pact was interrupted only by the election of Hugo Chávez in 1998.
7. The term 'sociedad abigarrada' was coined by Bolivian intellectual René Zavaleta Mercado (1935–1984) to refer to his native country where different ethnic groups do not mingle (Zavaleta Mercado, 1982).
8. 'Sumak kawsay' (also 'sumaq qamaña') refers to an heterogeneous set of conceptions that reclaim indigenous values as a counterpoint to market relations. It has played an important political role in recent years, particularly in Bolivia and Ecuador.
9. Venezuela's political developments under Chávez have been described as the 'Bolivarian Revolution', evoking South America's independence leader Simón Bolívar (1783–1830) who nurtured projects of regional integration.

Disclosure statement

No potential conflict of interest was reported by the author.

Funding

I thank the Fundação de Amparo à Pesquisa do Estado de São Paulo (FAPESP) for its support for this research [grant number 2014/05549-3].

References

A Fundação Alexandre de Gusmão. (2012). *América do Sul e a integração regional*. Brasília: Author.
Amorim, C. (2004). Conceitos e estratégias da diplomacia do governo Lula. *DEP- Diplomacia, Estratégia e Política. Ano, 1*(1), 41–48.
Bonomo, M., Brito, R., & Martins, B. (2014). *Macroeconomic and financial consequences of the after crisis government-driven credit expansion in Brazil* (Working Paper 378). Rio de Janeiro: Banco Central do Brasil.
Campos, P. H. (2009). As origens da internacionalização das empresas de engenharia brasileiras. In F. R. Luxemburgo (Ed.), *Empresas transnacionais brasileiras na América Latina: um debate necessário* (pp. 103–114). São Paulo: Expressão Popular.
Cervo, A. L. (2003). Política exterior e relações internacionais do Brasil: enfoque paradigmático. *Revista Brasileira de Política Internacional, 46*(2), 5–25.

Comissão Econômica para a América Latina e o Caribe. (1994). *El regionalismo abierto en América Latina y el Caribe: La integración económica al servicio de la transformación productiva con equidad*. Santiago de Chile: Comision economica para America Latina z el Caribe.

Consejo Suramericano de Infraestructura z Planeamiento. (2014, December 4). *Cartera de Proyectos 2014*. Montevideo: V Reunión ordinaria del COSIPLAN. Retrieved from http://www.iirsa.org/admin_iirsa_web/Uploads/Documents/cn25_montevideo14_Cartera_COSIPLAN_2014.pdf

Dos Santos, F. L. B. (2014). A problemática brasiguaia e os dilemas da influéncia regional brasileira. In W. A. D. Neto (Ed.), *O Brasil e novas dimensões da integração regional* (pp. 447–478). Rio de Janeiro: IPEA.

Drummond, C. (2014, May 19). Entrevista – Vera Thorstensen: 'Ficar atrelado ao Mercosul é afundar o Brasil'. Carta Capital. Retrieved from http://www.cartacapital.com.br/economia/ficar-atrelado-ao-mercosul-e-afundar-o-brasil-804.html

Fiori, J. L. (2011). A miséria do 'novo desenvolvimentismo'. Retrieved from https://www.cartamaior.com.br/?/Coluna/A-miseria-do-novo-desenvolvimentismo-/20887

Fundaçao Dom Cabral. (2010). *Ranking FDC das Transnacionais Brasileiras*. São Paulo: Fundaçao Dom Cabral.

Furtado, C. (1974). *O mito do desenvolvimento econômico*. Rio de Janeiro: Paz e Terra.

Garcia, A. E. S. (2012). A internacionalização de empresas brasileiras durante o governo Lula: uma análise crítica da relação entre capital e Estado no Brasil contemporâneo (Unpublished doctoral dissertation). Pontifícia Universidade Católica, Rio de Janeiro.

Garzon, L. F. N. (2013, May 3). A esfinge, o BNDES e as 'campeãs' que nos devoram. Correio cidadania. Retrieved from http://www.correiocidadania.com.br/index.php?option=com_content&view=article&id=8329%3Amanchete030513&catid=34%3Amanchete&amp

Gonçalves, J. B. (2013). Desafios da inserção internacional do Brasil: próximos passos. In L. Paz (Ed.), *O CEBRI e as relações internacionais no brasil*. São Paulo: Editora SENAC.

Initiative for the Integration of the Regional Infrastructure of South America. (2011). *IIRSA: 10 anos depois: Suas conquistas e desafíos*. Buenos Aires: BID-INTAL.

Lampréia, L. F. (2013). Os desafios do Brasil. In L. Paz (Ed.), *O CEBRI e as relações internacionais no Brasil*. São Paulo: Senac.

Leopoldo, R. (2011, March 10). *BNDES empresta 391% mais em 5 anos e supera em três vezes o Banco Mundial*. Estado de São Paulo. Retrieved from http://economia.estadao.com.br/noticias/geral,bndes-empresta-391-mais-em-5-anos-e-supera-em-tres-vezes-o-banco-mundial-imp-,689817

Luce, M. S. (2007). *O subimperialismo brasileiro revisitado: a política de integração regional do governo Lula (2003–2007)*. (Unpublished Master's dissertation). Universidade Federal do Rio Grande do Sul, Porto Alegre.

Mercadante, A. (2010). *As bases do novo desenvolvimentismo: análise do governo Lula*. (Unpublished doctoral thesis). Instituto de Economia da Universidade Estadual de Campinas.

Paulani, L. (2008). *Brasil delivery: Servidao financier e estado de emergência econômico*. São Paulo: Boitempo.

Pecequilo, C. S. (2008). A política externa do Brasil no século XXI: os eixos combinados de política externa do governo Lula (2003–2006). *Revista Brasileira de Política Internacional, 51*(2), 136–153.

Pinto, L., & Reis, M. (2017). Long-term finance in Brazil: The role of the Brazilian development bank (BNDES). In E. Grivoyannis (Ed.), *The New Brazilian economy* (pp. 151–176). New York: Palgrave Macmillan.

Rice, C. (2008, March 13). Remarks with Brazilian Foreign Minister Celso Amorim. U.S. Department of State. Retrieved from http://2001-2009.state.gov/secretary/rm/2008/03/102228.htm

Rodrigues, E. (2009, September 27). Brasil faz obras nos vizinhos temendo a China. *Folha de São Paulo*. Retrieved from http://www1.folha.uol.com.br/fsp/dinheiro/fi2709200910.htm

Rodrigues, A. (2010, February 15). BNDES aposta R$ 7,5 bi no Friboi: Concorrentes foram ao banco reclamar de privilégos. Estado de São Paulo. Retrieved from http://www.estadao.com.br/noticias/geral,bndes-aposta-r-7-5-bi-no-friboi,511466

Sader, E. (2010). *A nova toupeira e os caminhos da esquerda latino-americana*. São Paulo: Boitempo.

Sampaio, P. de. A. Jr. (2012). Desenvolvimentismo e neodesenvolvimentismo: tragédia e farsa. *Serviço Social & Sociedade, 112*. Online publication. Retrieved from doi:10.1590/S0101-66282012000400004

Serbin, A., Martínez, L., & Ramanzani Júnior, H. (2012). *El regionalismo 'post–liberal' en América Latina y el Caribe: Nuevos actores, nuevos temas, nuevos desafíos; Anuario de la Integración Regional de América Latina y el Gran Caribe*. Buenos Aires: Coordinadora Regional de Investigaciones Económias y Sociales.

Sherman, W. (2012, February 28). Under Secretary for Political Affairs. Remarks to the Council of the Americas and the Centre for Strategic and International Studies (CSIS). Carnegie Endowment for International Peace Washington, DC. Retreived from http://www.state.gov/p/us/rm/2012/184853.htm

Sicsú, J., de Paula, L. F., & Michel, R. (2005). *Novo-desenvolvimentismo: Um projeto nacional de crescimento com eqüidade social*. São Paulo: Manole.

Tautz, C., Siston, F., Pinto, J. R. L., & Badin, L. (2010). O BNDES e a reorganização do capitalismo brasileiro: um debate necessário. In *Os anos lula: Contribuições para um balanço crítico, 2003-2010* (pp. 249–286). Rio de Janeiro: Garamond.

Teixeira, C. G. P. (2011). Brazil and the institutionalization of South America: From hemispheric estrangement to cooperative hegemony. *Revista Brasileira de Política Internacional, 54*(2), 189–211.

Teixeira, R. A., & Neto, W. A. D. (2012). La recuperación del desarrollismo en el regionalismo Latinoamericano. In W. A. D. Neto & R. A. Teixeira (Eds.), *Perspectivas para la integración de América Latina* (pp. 11–36). Brasília: CAF and IPEA.

Vigevani, T., & Cepaluni, G. (2007). A política externa de Lula da Silva: a estratégia da autonomia pela diversificação. *Contexto Internacional, 29*(2), Rio de Janeiro, 273–335. doi:10.1590/S0102-85292007000200002

Visentini, P. F. (2008). *Relações internacionais do Brasil – De Vargas a Lula*. São Paulo: Fundação Perseu Abramo.

Zavaleta Mercado, R. (1982). *Clases sociales y conocimiento*. La Paz: Los Amigos del Libro.

Index

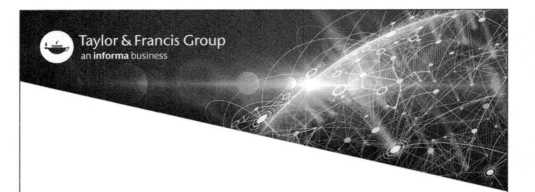

Printed in Great Britain
by Amazon

42867984R00071